Cybersecurity of Industrial Systems

Series Editor
Jean-Paul Bourrières

Cybersecurity of Industrial Systems

Jean-Marie Flaus

WILEY

First published 2019 in Great Britain and the United States by ISTE Ltd and John Wiley & Sons, Inc.

Apart from any fair dealing for the purposes of research or private study, or criticism or review, as permitted under the Copyright, Designs and Patents Act 1988, this publication may only be reproduced, stored or transmitted, in any form or by any means, with the prior permission in writing of the publishers, or in the case of reprographic reproduction in accordance with the terms and licenses issued by the CLA. Enquiries concerning reproduction outside these terms should be sent to the publishers at the undermentioned address:

ISTE Ltd
27-37 St George's Road
London SW19 4EU
UK

www.iste.co.uk

John Wiley & Sons, Inc.
111 River Street
Hoboken, NJ 07030
USA

www.wiley.com

© ISTE Ltd 2019
The rights of Jean-Marie Flaus to be identified as the author of this work have been asserted by him in accordance with the Copyright, Designs and Patents Act 1988.

Library of Congress Control Number: 2019939692

British Library Cataloguing-in-Publication Data
A CIP record for this book is available from the British Library
ISBN 978-1-78630-421-6

Contents

Foreword xiii

Introduction xix

Chapter 1. Components of an Industrial Control System 1

   1.1. Introduction. 1
      1.1.1. Definition: automated and cyber-physical systems 1
      1.1.2. Definition: Information System (IS). 1
      1.1.3. Definition: industrial IS or ICS. 2
      1.1.4. Definition: IT and OT system. 4
      1.1.5. Definition: SCADA 4
      1.1.6. Definition: Distributed Control Systems (DCS). 5
      1.1.7. Definition: Industrial Internet of Things (IIOT). 5
      1.1.8. Different types of ICS 6
   1.2. From the birth of the PLC to the SCADA system 6
   1.3. Programmable logic controller (PLC) 8
   1.4. RTU, master terminal unit and intelligent electronic device. 12
   1.5. Programmable Automation Controller 13
   1.6. Industrial PC 13
   1.7. Safety instrumented systems. 13
   1.8. Human–machine interface (HMI). 15
   1.9. Historians. 17
   1.10. Programming and parameter setting stations 17
   1.11. Industrial Internet of Things (IIoT) 18
   1.12. Network equipment 19
      1.12.1. Switch and hub 19
      1.12.2. Router and gateway. 20
      1.12.3. Firewall. 20
      1.12.4. IoT gateway 20

1.13. Data processing platform 21
1.14. Lifecycle of an ICS 22

Chapter 2. Architecture and Communication in an Industrial Control System 25

2.1. Network architecture 25
   2.1.1. Purdue model and CIM model 26
   2.1.2. Architecture of the Industrial Internet of Things 29
2.2. Different types of communication networks 31
   2.2.1. Topology 31
   2.2.2. Types of networks 33
   2.2.3. Virtual private network 34
   2.2.4. OSI model 34
2.3. Transport networks 35
   2.3.1. Ethernet 35
   2.3.2. Wi-Fi 36
   2.3.3. The IEEE 802.15.1 (Bluetooth) standard 36
   2.3.4. IEEE 802.15.4 networks 37
   2.3.5. LPWAN networks 38
   2.3.6. Cellular networks 38
2.4. Internet protocols 39
   2.4.1. The Internet protocol 39
   2.4.2. Transmission Control Protocol 39
   2.4.3. Unified Datagram Protocol (UDP) 42
   2.4.4. Address Resolution Protocol (ARP) 42
   2.4.5. Internet Control Message Protocol (ICMP) 42
   2.4.6. The IPv6 protocol 43
2.5. Industrial protocols 43
   2.5.1. Introduction 43
   2.5.2. Modbus 45
   2.5.3. Profibus and Profinet 46
   2.5.4. Actuator/sensor interface 47
   2.5.5. Highway Addressable Remote Transducer 48
   2.5.6. DNP3 and IEC 60870 48
   2.5.7. The CAN bus 49
   2.5.8. Ethernet/IP and Common Industrial Protocol (CIP) 49
   2.5.9. OLE for Process Control (OPC) 51
   2.5.10. Other protocols 52
2.6. IoT protocols 52
   2.6.1. 6LowPAN 53
   2.6.2. Message Queuing Telemetry Transport 53
   2.6.3. CoAP 54
   2.6.4. Other protocols 54

Chapter 3. IT Security ... 57

3.1. Security objectives ... 57
    3.1.1. The AIC criteria ... 57
    3.1.2. The different levels of IT security ... 61
3.2. Differences between IT and OT systems ... 64
    3.2.1. The functionalities ... 64
    3.2.2. The technology ... 65
    3.2.3. System lifecycle ... 66
    3.2.4. Security management ... 67
    3.2.5. IT/OT convergence ... 68
    3.2.6. Summary ... 68
3.3. Risk components ... 70
    3.3.1. Asset and impact ... 70
    3.3.2. Threats ... 71
    3.3.3. Attacks ... 71
    3.3.4. Vulnerabilities ... 72
    3.3.5. Definition of risk ... 73
    3.3.6. Scenarios and impact ... 74
    3.3.7. Risk measurement ... 75
3.4. Risk analysis and treatment process ... 77
    3.4.1. Principle ... 77
    3.4.2. Acceptance of risk ... 79
    3.4.3. Risk reduction ... 79
3.5. Principle of defense in depth ... 80
3.6. IT security management ... 82
3.7. Risk treatment process ... 85
3.8. Governance and security policy for IT systems ... 86
    3.8.1. Governance ... 86
    3.8.2. Security policy ... 87
3.9. Security management of industrial systems ... 88

Chapter 4. Threats and Attacks to ICS ... 91

4.1. General principle of an attack ... 91
4.2. Sources of threats ... 95
4.3. Attack vectors ... 98
4.4. Main categories of malware ... 99
    4.4.1. Virus/worms ... 100
    4.4.2. Trojan horse ... 100
    4.4.3. Logical bomb ... 101
    4.4.4. Rootkit ... 101
    4.4.5. Spyware ... 101
    4.4.6. Back doors ... 101

4.4.7. Botnet	102
4.4.8. Ransomware	103
4.5. Attacks on equipment and applications	103
4.5.1. Buffer overflow and integer overflow	103
4.5.2. Attack by brute force	104
4.5.3. Attack via a zero day flaw	105
4.5.4. Side-channel attacks	105
4.5.5. Attacks specific to ICS equipment	106
4.5.6. Attacks on IIoT systems	107
4.6. Site attacks and via websites	108
4.7. Network attacks	109
4.7.1. Man-in-the-middle	109
4.7.2. Denial of service	110
4.7.3. Network and port scanning	111
4.7.4. Replay attack	112
4.8. Physical attacks	112
4.9. Attacks using the human factor	113
4.9.1. Social engineering	113
4.9.2. Internal fraud	114
4.10. History of attacks on ICS	114
4.11. Some statistics	119

Chapter 5. Vulnerabilities of ICS 121

5.1. Introduction	121
5.2. Generic approach to vulnerability research	122
5.3. Attack surface	124
5.4. Vulnerabilities of SCADA industrial systems	126
5.5. Vulnerabilities of IoT industrial systems	128
5.6. Systematic analysis of vulnerabilities	130
5.7. Practical tools to analyze technical vulnerability	136
5.7.1. Databases and information sources	137
5.7.2. Pentest tools	137
5.7.3. Search engines	139

Chapter 6. Standards, Guides and Regulatory Aspects 141

6.1. Introduction	141
6.2. ISO 27000 family	142
6.3. NIST framework and guides	144
6.3.1. NIST Cyber Security Framework	144
6.3.2. The guides	145
6.4. Distribution and production of electrical energy	148
6.4.1. NERC CIP	148

6.4.2. IEC 62351 150
6.4.3. IEEE 1686 151
6.5. Nuclear industry 151
6.5.1. The IAEA technical guide. 151
6.5.2. IEC 62645 152
6.6. Transportation 153
6.6.1. Vehicles 153
6.6.2. Aeronautics 153
6.7. Other standards. 154
6.7.1. National Information Security Standards 154
6.7.2. Operating safety standards 154
6.8. ANSSI's approach. 155
6.9. Good practices for securing industrial Internet of Things equipment . . 159
6.9.1. Trust base (root of trust) 160
6.9.2. Identity management (endpoint identity) 161
6.9.3. Secure boot 161
6.9.4. Cryptographic services. 161
6.9.5. Secure communications 162
6.9.6. Equipment configuration and management. 162
6.9.7. Activity dashboard and event management by a SIEM. 162
6.10. Legislative and regulatory aspects. 163

Chapter 7. The Approach Proposed by Standard 62443 167

7.1. Presentation. 167
7.2. IACS lifecycle and security stakeholders 169
7.3. Structure of the IEC 62443 standard 170
7.4. General idea of the proposed approach 172
7.5. Basics of the standard. 174
7.5.1. Fundamental requirements 174
7.5.2. Security Levels (SL) 177
7.5.3. Zones and conduits 180
7.5.4. Maturity level 182
7.5.5. Protection level 183
7.6. Risk analysis 184
7.6.1. General approach 185
7.6.2. Detailed risk analysis. 186
7.6.3. Determination of SL-T. 187
7.6.4. Countermeasures 188
7.7. Security management 189
7.8. Assessment of the level of protection 190
7.9. Implementation of the IEC 62443 standard 191

        7.9.1. Certification   191
        7.9.2. Service providers and integrators   192
        7.9.3. IACS Operators   192

Chapter 8. Functional Safety and Cybersecurity   193

    8.1. Introduction.   193
        8.1.1. Components of operational safety   193
        8.1.2. SIS and SIL levels   198
    8.2. IEC 61508 standard and its derivatives   200
    8.3. Alignment of safety and security   203
    8.4. Risk analysis methods used in operational safety   204
        8.4.1. Preliminary hazard analysis.   204
        8.4.2. Failure Mode and Effects Analysis   205
        8.4.3. HAZOP.   207
        8.4.4. Layer Of Protection Analysis.   208
        8.4.5. Fault trees and bowtie diagrams   210

Chapter 9. Risk Assessment Methods   213

    9.1. Introduction.   213
    9.2. General principle of a risk analysis   214
        9.2.1. General information   214
        9.2.2. Setting the context   217
        9.2.3. Risk identification   218
        9.2.4. Estimation of the level of risk   219
        9.2.5. Risk assessment and treatment   219
        9.2.6. Tailor-made approach and ICS   221
    9.3. EBIOS method   221
        9.3.1. Workshop 1: framing and security base.   222
        9.3.2. Workshop 2: sources of risk   226
        9.3.3. Workshop 3: study of strategic scenarios   227
        9.3.4. Workshop 4: study of operational scenarios   229
        9.3.5. Workshop 5: risk treatment   230
        9.3.6. Implementation for ICS   233
    9.4. Attack trees   234
    9.5. Cyber PHA and cyber HAZOP   236
        9.5.1. Principle   236
        9.5.2. Cyber PHA.   239
        9.5.3. Cyber HAZOP   243
    9.6. Bowtie cyber diagram.   245
    9.7. Risk analysis of IIoT systems   246

Chapter 10. Methods and Tools to Secure ICS 249

10.1. Identification of assets. 249
10.2. Architecture security. 253
   10.2.1. Presentation 253
   10.2.2. Secure architecture 254
   10.2.3. Partitioning into zones 255
10.3. Firewall 257
10.4. Data diode. 260
10.5. Intrusion detection system 261
   10.5.1. Principle of operation. 261
   10.5.2. Detection methods 264
   10.5.3. Intrusion detection based on a process model 267
10.6. Security incident and event monitoring. 268
10.7. Secure element 270

Chapter 11. Implementation of the ICS Cybersecurity Management Approach 273

11.1. Introduction. 273
   11.1.1. Organization of the process 273
   11.1.2. Technical, human and organizational aspects 275
   11.1.3. Different levels of implementation and maturity. 275
11.2. Simplified process 276
11.3. Detailed approach 277
11.4. Inventory of assets 279
   11.4.1. Mapping 279
   11.4.2. Documentation management 279
11.5. Risk assessment 280
11.6. Governance and ISMS 281
   11.6.1. Governance of the ICS and its enviroment 281
   11.6.2. ISMS for ICS 281
11.7. Definition of the security policy and procedures 282
11.8. Securing human aspects. 283
11.9. Physical security 284
11.10. Network security 285
11.11. Securing exchanges by removable media. 285
11.12. Securing machines 285
   11.12.1. Securing workstations and servers 285
   11.12.2. Securing engineering stations 286
   11.12.3. Securing PLCs 286
   11.12.4. Securing IIoT equipment 287
   11.12.5. Securing network equipment. 287
   11.12.6. Antivirus 287

11.13. Data security and configuration	288
11.14. Securing logical accesses	289
11.15. Securing supplier and service provider interactions	290
11.16. Incident detection	291
11.16.1. Logging and alerts	291
11.16.2. Intrusion detection system	291
11.16.3. Centralization of events (SIEM)	291
11.17. Security monitoring	291
11.17.1. Updating mapping and documentation	291
11.17.2. Security patch management	291
11.17.3. Audit of the facility	292
11.18. Incident handling	292
11.19. Recovery	293
11.19.1. Backup	293
11.19.2. Business continuity plan	294
11.20. Cybersecurity and lifecycle	294
Appendix 1	295
Appendix 2	303
Appendix 3	309
Appendix 4	329
Appendix 5	355
Appendix 6	361
List of acronyms and abbreviations	363
References	367
Index	377

Foreword

Cybersecurity is one of the major concerns of our time. The risk of cyber-attacks accompanies the development of digital systems and their networking, particularly through the Internet. These risks concern all types of installations, and attacks can be carried out by isolated actors – "hackers" who, depending on their ethics (provided they have any) will be qualified as "white, grey or black hats" – but they can also be the work of international criminal organizations, or even State services acting at offensive or counteroffensive level.

The motivations for these attacks are very diverse: the desire to disrupt, harm or even destroy, theft of information, threats, intimidation, blackmail, revenge, extortion, demonstration of force, etc. There are now countless examples of this, and industrial systems, small or large, which were long thought to be protected because of their specific characteristics and their isolation from the outside world (the famous air gap), are no longer immune to threats of very different shapes and sizes.

The consequences of successful attacks can be serious because in the industrial world, the aim will of course be to protect the information system and the data it contains, but the primary objective is to prevent serious disruptions in controlled processes. These disruptions can lead to untenable production stoppages for manufacturers, regardless of their size, and generate damage to the environment, property and people, with consequences that can be major. It is easy to imagine disaster scenarios that could affect sensitive installations in the fields of energy production, water treatment, transport and more generally major infrastructure.

The industry therefore faces a real problem that it can no longer ignore and it is the duty of each manager to assess the risks to which the installation for which he/she is responsible is exposed, and to take appropriate protective measures. However, industrial managers remain perplexed about the measures to be taken and the organization to be put in place. If they are willing to acknowledge the reality of risk, they often have difficulty perceiving its origin and magnitude, and admitting its possible consequences.

Yet, for a long time, the industry has been accustomed to dealing with functional safety and the risks of component and component failure, and operator manipulation errors that can affect essential functionality. The understanding of these risks has given rise to international standards: IEC 61508 on the functional safety of electrical/electronic/electronic programmable systems, and IEC 61511 specific to the processing industries sector, itself based on the ISA-84 standard developed by the ISA (International Society of Automation). These problems can be addressed probabilistically from experimental and experiential data, as the threats are unintentional.

In the case of cybersecurity, we know that there are threats, which will come from the outside, perhaps also from the inside, but in what form, with what magnitude and with what probability? In the case of threats of intentional actions, this is a purely subjective area of assessment that can lead to an overestimation, resulting in a level of protection that will be detrimental to the company's competitiveness, or an underestimation that will pose an intolerable risk to the company.

In addition, attack techniques are evolving and improving. From the simple viruses of the 1990s, detectable by their signature, we have moved on to malicious software, which are complex computer constructions capable of communicating with the outside world, capable of growing and becoming more widespread, and capable of taking remote control of installations. Some attacks are targeted, as were the attacks on Ukrainian power grids in late 2015 and 2016, while others are broad spectrum, such as the Wannacrypt and NotPetya attacks, which have caused serious disruptions on many industrial installations, including in France.

Companies can be held for ransom from ransomware that has become common practice; they can also be complicit without their knowledge in distributed denial-of-service attacks, because connected objects – especially those that are permanently connected to the Internet but are insufficiently protected: surveillance cameras, printers, boxes – can be enrolled in botnets, manipulated at a distance to participate in massive attacks.

The development of the industrial Internet of Things will greatly expand the attack surfaces with the networking of a considerable number of diverse devices that will be impossible to monitor individually, and from which we will have to be wary of the origin, development conditions and the way they store and exchange information.

People working in the industrial world are often confused about how to approach the problem, but the normative and regulatory context forces them not to remain inactive. In France, ANSSI was charged by the Military Programming Act of December 18, 2013 and the decrees of March 27, 2015, with ensuring the security of vital operators' information systems. More recently, the European *Network and Information Security* (NIS) directive, transposed into French law by the law of February 26, 2018 and the decree of May 23, 2018, introduced obligations for all operators of essential services.

It is likely that insurers will also exert increasing pressure for all companies to take appropriate protective measures.

Jean-Marie Flaus' book is therefore timely and meets an essential need. It is an extremely valuable tool to better understand cybersecurity issues and solutions. Jean-Marie Flaus is a professor at the University of Grenoble Alpes. He is also a teacher-researcher and head of the Department of Management and Control of Production Systems at the G-SCOP Laboratory, Science for Design, Optimization and Production. The laboratory G-SCOP is a multidisciplinary laboratory created in Grenoble in 2007 by the CNRS, Grenoble-INP and the University of Grenoble Alpes, in order to meet the scientific challenges posed by changes to the industrial sector. Cybersecurity is clearly one of them.

The author addresses it in his book with both a teacher's and a practitioner's eye. His approach is deliberately didactic and aims to provide a detailed understanding of the nature and extent of the threats facing the industry. Its purpose is not to alarm unnecessarily, but to provide the keys to an assessment that is as objective as possible of the risks involved, which will be collated with those that a functional safety analysis may have revealed in order to identify the industrial risks involved as completely and as homogeneously as possible.

But Jean-Marie Flaus is also a practitioner, leading in particular the work of the "Cybersecurity of industrial installations and the Internet of Things" group within the Institute for Risk Management (IMdR). Once the overview of threats and vulnerabilities has been established, the author outlines the approach to be followed to address them based, in particular, on the normative standards that can be used. The fabric of standards is often considered complex and abstruse but, without getting lost in their mysteries, Jean-Marie Flaus explains its philosophy and approach, focusing on the two most important ones: the ISO 27000 series of standards and the IEC 62443 series. This last set of standards is the result of a long process of work undertaken within the ISA99 committee of the ISA more than 10 years ago and now in the process of being completed. The IEC 62443 standard is the only normative text specifically dedicated to industrial control systems; it has a double merit:

– on the one hand, it segregates the obligations to be met throughout the lifecycle of a control system according to the role played: product developer or manufacturer, integration service provider, operator, maintenance service provider;

– on the other hand, it provides the link and synthesis between the technical and organizational measures necessary to achieve a given level of security following a risk analysis.

As Jean-Marie Flaus explains very well, organizational and technical aspects must go hand in hand. There is no point in installing firewalls if the way they are operated and programmed is not defined. Conversely, "policies & procedures", as sophisticated as they may be, are of no interest if they are not technically supported.

The reader will find in the book a description of the traditional and most advanced protection techniques, but also a statement of the rules and method to be followed to build an information security management system adapted to the case of each industrial installation. For such a system to be complete, it is necessary to think in terms of "protection" but also to act at the level of "prevention" and "early detection" of intrusions, in particular abnormal traffic suggesting that an attack is in preparation. It is also necessary, because the hypothesis of a successful attack cannot be ruled out, to consider how to contain it, through appropriate defense in depth, and to restore the normal functioning of the system, starting with essential services.

Jean-Marie Flaus makes a clear and precise presentation of all this, without ever falling into abstraction, and also dealing with a simplified approach to risk management, when the stakes are low and do not justify overly sophisticated analyses.

It is a book from which certain chapters can be extracted for a thorough reading; it is also a book that can be read in its entirety, without boredom and where much is learned. It is certainly a book that will become a reference work that each manufacturer must have *at least* consulted and kept nearby, and that will be extremely valuable to all cybersecurity professionals to better understand the issues and solutions.

Jean-Pierre HAUET
President of ISA-France
Voting member of the ISA99 committee

Introduction

I.1. What is industrial cybersecurity?

Nowadays, more and more man-made physical systems are controlled by a computer system. This is the case for autonomous systems such as vehicles, everyday devices and industrial production systems or water or energy distribution systems. Most of these systems are also connected in some way to the Internet.

The computer security of this equipment is becoming a major issue for the industrial world. This is particularly true today, in the context of the factory of the future, also known as Industry 4.0, which is presented as the fourth industrial revolution, and which is characterized by increasingly connected systems and by the increasingly strong integration of digital technologies into manufacturing processes.

There are many spectacular cyber-attacks in the news: they aim to steal identifiers, make some systems or websites unable to function properly, or try to block workstations by encrypting data in order to obtain a ransom.

The control systems of industrial installations are also subject to attacks, either by the collateral effect of a computer attack, as in the case of WannaCry (Symantec 2017; May *et al.* 2018), which can lead to plant shutdown and significant operating losses, or specifically with an attack on industrial systems. This is the case with the Stuxnet attack (Falliere *et al.* 2011), which aimed to destroy uranium production capacities in Iran, or the Triton attack (White 2017), which aimed to render security systems inoperative. Other recent attacks are presented in Chapter 4.

The potential damage that can be caused is great and ranges from simple yield losses to material and human damage, as well as information losses, which can be very serious. Given the magnitude of the impact of this potential damage, and given the frequency of attacks, the risk associated with the cybersecurity of industrial systems has become significant.

For a long time, this risk was neglected: industrial installations were loosely connected to company networks or the Internet, and industrial control system (ICS) seemed to be protected. The evolution of technology, uses and needs has led to the connection of these systems to other networks, whether for the transfer of production data to the company's IT systems, for remote maintenance or for automatic download of updates. At the same time, the convergence of protocols toward common standards has increased the vulnerability of control systems. The idea that industrial systems could be considered isolated from the rest of the world, sometimes referred to as the air gap myth, is unrealistic nowadays.

The aggravating factor is that, since most technologies and protocols were developed at a time when cyber-attacks did not exist, they are not very secure and highly vulnerable. Many installations still use them. As the renewal rate of systems and equipment used in ICS is very low, relatively old equipment is still in service. The lifetime of the installations, which is much longer for ICS than for traditional computer systems, is an additional vulnerability. For the most recent installations that are being implemented around the industrial Internet of Things, a new risk factor is emerging, which is related to the complexity and flexibility of the installations.

The risk is therefore very real and cannot be ignored. It must therefore be controlled. As in other areas, there is no such thing as zero risk, so we must ensure that it is contained at an acceptable level. Therefore, the risk must be assessed and dealt with appropriately, that is, select the relevant actions to be taken. Depending on the issues and the context, the answers will of course be different. Measures cannot be limited to technical actions, they must be part of a risk management plan (Chapter 3) that can remain simple, but which must be global and take into account human and organizational aspects. Moreover, since, as we said above, there is no such thing as zero risk, it is useful to set up a recovery and business continuity plan, or even a crisis management plan. This will be based on a detection system and an alert chain. The whole approach must of course take into account the cost–benefit ratio.

I.2. From information security to cybersecurity

The security of an Information System (IS) concerns all aspects related to the control of IS risks and aims to guarantee:

– the optimal functioning of the IS to obtain the best *quality of service*;

– that no unacceptable damage may affect the various elements of the IS (beyond a fixed level);

– that no undesired operation may lead, directly or indirectly, to unacceptable damage to the rest of the company or partners (beyond a fixed level).

The term "cyber" is a prefix from the Greek word *Kubernêtikê* meaning "to lead, to govern". In 1948, Norman Wiener introduced the term "cybernetics" to refer to the sciences of control and communication between the living being and the machine. "Cyber" has become relative to what is related to computers, and we speak of cyberspace to refer to the extension of our natural space through the Internet.

Cybersecurity concerns the computer security of systems connected to the Internet and belonging to cyberspace. Cyber-attacks are computer attacks in cyberspace, in addition to existing threats to ISs.

Through abuse of language, we often speak of cybersecurity for everything related to computer security (Niekerk and Solms 2016).

I.3. Is there really a risk for industrial systems?

Cyber-attacks are not very relevant to industrial or cyber-physical systems.

It is true that most attacks concern traditional computer systems. These attacks are endless and the tools to create attacks are becoming more democratic. The means used by organized cybercrime have grown considerably.

As far as industrial systems are concerned, attacks are limited in number and often show that the attacker has a very specific knowledge of the systems under attack and has implemented a tailor-made attack.

Does this mean that the risk associated with the cybersecurity of industrial systems is low? The answer is of course no. The level of risk depends on the severity of the damage and the likelihood of its occurrence. For an industrial or nuclear installation, the damage can be catastrophic and impact the population. The possibility of damage being caused is at least at the same level as for IT management systems. The level of risk is therefore very high. To be convinced, it is sufficient to observe the evolution of regulatory obligations, such as the LPM (Military Programming Act) in France, the NIS (Network and Information Security) directive in Europe or the Critical Infrastructures Protection Act in the United States.

The system is isolated from the Internet, so it is safe.

For a long time, it was believed that not being connected to the Internet was enough to avoid any risk of computer piracy. We sometimes talk about the myth of the air gap. In fact, the situation is more complex:

– first of all, even if a system is not connected, it can be a victim of technological malicious acts, Stuxnet is a demonstrative example. The attack vector was a USB key;

– often, the industrial network is connected to the IT corporate network: this IT system can be the victim of attacks and host malicious programs that subsequently attempt to corrupt the industrial network, or it can even allow attacks to pass directly through;

– in an industrial network, there are sometimes direct connections to the Internet, more or less official and sometimes temporary, for maintenance or configuration, and these represent a real vulnerability.

In addition, with the growing need to upload data to the IS or to the Cloud, with update systems from a manufacturer's site and with remote maintenance, the isolation of industrial systems is increasingly illusory.

The stakes are low.

A widespread idea is that the risk is low when the production equipment does not use dangerous machines or processes.

For this type of installation, it is clear that the damage to the environment and people will be limited. However, for the company, the impact can be enormous since an attack can result in a shutdown of production for a long period of time, a substandard quality of the products manufactured, or even a destruction of the production equipment. The economic consequences must be analyzed and a cost–benefit analysis carried out to determine the level of cybersecurity measures to be taken.

The workstations are equipped with antivirus software and there is a firewall, and therefore we are protected.

Using an antivirus is a basic step to take. It allows the protection of computer workstations, running under Windows or macOS. However, in an industrial system, there are many devices running on a real-time operating system or an embedded Linux system, or even a proprietary system. For these systems, there is not necessarily an antivirus and they are therefore vulnerable.

Let us add that one of the problems with antivirus software for industrial computer workstations is that they are not always updated.

The limits of firewalls are also well known: the first is that filtering rules are not always well configured; the second is that even if data flows are limited, this does not prevent all attacks from passing through. As detailed in Chapter 4, the electrical energy management system that attacked in Ukraine in 2015 included firewalls that did not prevent anything from happening.

We use a Virtual Private Network (VPN), so no problem.

Another widespread idea is that the use of a VPN provides effective protection. This idea is also wrong for two main reasons:

– the first is that a significant number of VPNs use technologies considered outdated and are therefore vulnerable. A 2016 study (High-Tech Bridge Security Research 2016) showed that 77% of the VPNs tested still use a SSLv3-based protocol, created in 1996, or even SSLv2, while most standards such as PCI DSS or NIST SP 800-52 (see Chapter 6) prohibit its use;

– the second is that even with a well-configured VPN, if one of the workstations connected to these VPNs has been compromised, it can compromise the rest of the network, all the more easily because the transported data are encrypted and therefore difficult to filter.

Information System Security (ISS) is expensive and generates many constraints that limit efficiency.

A common idea is that ISS is expensive and imposes a large number of operational constraints that are incompatible with those of industrial control systems.

These constraints appear all the more important as an ICS uses very heterogeneous equipment, and its users appreciate a certain flexibility for operation. For example, the use of mobile terminals is becoming more and more common, as it is very useful for system control close to the process.

In reality, the ISS of industrial systems must be adapted to the challenges, and it is important to carry out a risk analysis and to compare the importance of these risks with the cost of measures to reduce them and the constraints they impose. However, security is often considered a source of expenditure that is difficult to justify by a return on investment. It is more relevant to measure it against a potential loss, for example on the quantity of production, if the system is unavailable, or on the cost of reconstruction, if it is destroyed.

As for operating constraints, it is important to implement solutions in consultation with users and taking into account the reality on the ground. We must support new operating modes and not try to limit users' possibilities excessively.

I.4. Vulnerability of ICS

Computer systems are vulnerable to cyber-attacks and physical attacks. ICSs, which, as we have seen, are directly or indirectly connected to the Internet, are also vulnerable.

For historical reasons, and because of cultural differences, ICSs are even more vulnerable than traditional computer systems.

In general, an industrial installation is carried out for a relatively long period of time. Systems more than 10 years old are commonly found. In addition, the main objective is to continuously operate the production system, and everything that requires or can generate a shutdown is avoided. In addition, for accounting reasons, the protocols used are of a fairly old design and are not very secure.

But over time, ICSs have become highly connected. First, production and supply chain management processes are integrated with key management software, and production data are used in business computing applications. In addition, in order to allow a better reactivity, maintenance and monitoring are carried out remotely. Finally, some installations are isolated and supervised remotely (water treatment plant, energy distribution, etc.).

Industrial systems also have the specificity of being very heterogeneous and built from generic elements (Commercial Off-The-Shelf, COTS), chosen above all for their functionality and not for their security. The products used are not very secure, not always fully tested. Updating software, operating systems or firmware is difficult and is not always done regularly. Finally, at the equipment level, authentication management is difficult, and default passwords are not always changed.

Users are not always aware of the vulnerability of ICS: many physical ports (USB, RJ45) are poorly protected, and it may be easy to connect to them. There are even sometimes unofficial connections to the Internet, which have been created for reasons related to maintenance or remote access.

In addition, ICS are used in the world of production, where culture is often based on principles such as producing first and not taking the risk of changing something that works. This means, for example, that updates are not always made on a regular basis and that authentication management is not as rigorous as in the IT world with passwords that can be shared.

Finally, in many cases, there is no ICS security management policy: the management of subcontractors and stakeholders is not subject to specific measures, and there is no user access rights management policy defining their possibilities of action and prohibiting access by employees no longer belonging to the company.

I.5. Cybersecurity and functional security

Industrial installations and cyber-physical systems can be subject to failures or operating errors that can lead to harmful operations or failures. The study of these risks is the subject of what is called "functional security" or "operational safety".

The sources of dysfunction considered are numerous: they can be technical, human or organizational. In the technical origins, a category is the subject of particular attention, namely electrical, electronic or programmable electronic components. Standard 61508 (Chapter 8) describes the approach for analyzing the risks associated with their potential failures. The idea is to characterize the installation by a level of probability of failure over a given period of time. By coupling it with an impact analysis in the event of malfunctions, it is possible to assess the level of risk of an installation or system and to satisfy some given objectives.

Functional safety, as defined by IEC 61508, does not include *computer security*, which was not a risk identified as such in 2002, at the time the standard was written. Even the latest updates are only beginning to mention it. It should be noted that Information System Security (ISS) is also not explicitly included in risk analyses of hazardous installations called "hazard studies".

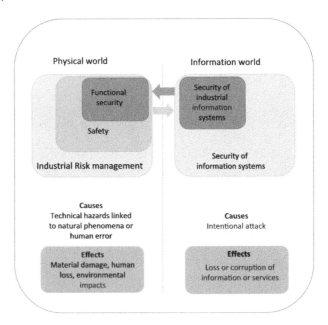

Figure I.1. *Relationship between functional security and information system security*

Information security and the operational safety of physical systems are managed by different approaches:

– on the one hand, the ISS risk management process is described in the 27000 family of standards (Chapter 6). The risk analysis methods used (Chapter 9) are, for example, EBIOS or OCTAVE methods;

– on the other hand, there are risk analysis methods for industrial processes such as PHA, HAZOP, FMEA or LOPA (Chapter 8) and standard 61508.

For ISS, the impact of the physical world on the IS, such as a power failure or overheating, is taken into account. On the other hand, the opposite influence (red arrow in figure I.1) is not considered for these systems, since they only handle information.

In the case of cyber-physical systems such as automated industrial systems or critical infrastructures, this influence exists and can have very significant damaging consequences. For these systems, it appears that the two aspects are not independent and must be addressed in a coordinated way.

Indeed, even if the causes of the feared events are different, the consequences and damage produced are for the most part common: damage to property, to people and to the environment through a process that is out of control. For example, a malfunction of a safety programmable logic controller (PLC) may be caused by both problems related to operational safety and to ISS. In any case, the consequences will be damage to the installation.

IEC 62443 (Chapter 7) aims to transcribe a number of functional safety concepts, including SIL levels, for cybersecurity in order to align approaches. For risk analysis, unified approaches are beginning to be proposed (Chapter 9).

I.6. Evolution of the perception of cybersecurity

The technological evolution that enabled the implementation of large-scale automated systems was the invention of the microprocessor by Intel in 1971. This led to the development of personal computing and, a few years later, industrial control systems. These systems were first connected by point-to-point wireline links, and then in the 1990s Ethernet communication with Transmission Control Protocol (TCP)/Internet Protocol (IP) became widespread. The main industrial protocols have been "encapsulated" and have allowed the development of industrial control systems as we know them.

At the time, the main problem was performance, and no one imagined that computer security would take on the proportions it has today.

The first computer viruses date back to the late 1980s. The brain virus that infects the PC "boot" sector was written in 1986 by brothers Basit and Amjad Farooq Alvi (Brain virus n.d.). It is recognized as the first virus on MS-DOS. In 1998, the Morris worm spreading on the Internet was written by R. Tappan. One of the first antiviruses (Norton antivirus) appeared in 1990. Virus progress was rapid, as it was also in 1990 that the first polymorphic virus appeared (Chamelon, written by R. Burger).

As early as the mid-2000s, the question of the vulnerability of ICS was raised (Abshier 2004; Wooldridge 2005). In 2008, the Aurora project (Meserve 2007), an experiment supervised by the Idaho National Laboratory, demonstrated the possibility of destroying an energy generator through a cyber-attack.

In 2010, a study on power generation systems conducted by Red Tiger, at the request of the U.S. Department of Homeland Security (Pollet 2010), showed that the IT security of these systems was not as par with that of the IT world. For example, many of the vulnerabilities made public were not fixed and left these systems vulnerable.

Since 2010, various regulations and bills have appeared in several countries. Let us quote:

– the National Cybersecurity and Critical Infrastructure Protection Act of 2013, USA;

– Article 22 of the Military Programming Act in 2013, France;

– the *IT Security Act* in 2016, Germany;

– the NIS Directive in Europe in 2016.

In addition, since 2010, a number of guides and books have been published on this subject (Macaulay and Singer 2012; Knapp and Thomas 2015), and methodological tools have been proposed to improve the ISS of industrial facilities (ANSSI 2012a; Stouffer *et al.* 2015).

Nevertheless, risk perception remained limited and steps were slow to be taken. However, with increasingly common and high-profile attacks, the consideration of cybersecurity in new projects is gradually emerging, and is becoming essential with the 2018 regulatory requirements (Chapter 6).

I.7. Structure of the book

This book is organized as follows[1]:

– Chapter 1 presents the different elements of an industrial control system, Supervisory Control And Data Acquisition (SCADA) and Industrial Internet of Things (IIoT);

– Chapter 2 describes the architecture of these systems and the different characteristics of the networks used in SCADA or IIoT systems;

– Chapter 3 presents the basic concepts of information security and risk management;

– Chapters 4 and 5 detail the principle of attacks and the analysis of ICS vulnerabilities;

– Chapters 6 and 7 present the standards and regulations, with one chapter dedicated to IEC 62443, which is the reference standard;

– Chapter 8 presents the useful concepts of operational safety;

– Chapter 9 focuses on risk analysis methods: the EBIOS method and some more specific methods for industrial systems: cyber APR, cyber HAZOP and cyber-bowtie;

– Chapter 10 presents methods and tools for securing an ICS: installation inventory, architecture security, and technical devices such as intrusion detection systems, data diodes and secure IIoT components. Cryptographic concepts are explained in Appendix 1 and the blockchain for IoT is introduced in Appendix 2;

– Chapter 11 proposes a comprehensive approach for ICS security. It is based on the standards and methods described above, and is presented in a simplified and detailed version.

---

1 Further information can be found at: http://industrial cybersecurity.io/.

1

Components of an Industrial Control System

1.1. Introduction

1.1.1. *Definition: automated and cyber-physical systems*

The systems we are interested in are automated systems in the broad sense: they consist of a computerized control system, sensors to access physical quantities and actuators to act on the controlled system.

Such systems are found in industrial production facilities, where control is carried out by a programmable logic controller (PLC) or by a set of connected objects. There are also such systems for the automation of buildings, water and electricity distribution networks, transport systems, etc. They have a computer part constituting an information system (IS) and a physical part. We are talking about a cyber-physical system (CPS).

1.1.2. *Definition: Information System (IS)*

An Information System (IS) is an organized set of resources (hardware, software, personnel, data, procedures, etc.) for acquiring, processing and storing information (in the form of data, text, images, sounds, etc.) within and between organizations.

Many ISs only process information, others affect the physical world. It is the latter that we are interested in this book.

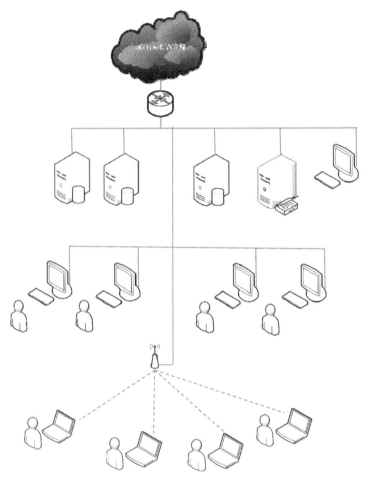

Figure 1.1. *Information system. For a color version of this figure, see www.iste.co.uk/flaus/cybersecurity.zip*

1.1.3. *Definition: industrial IS or ICS*

An industrial IS, or an industrial control system (ICS), is a system composed of an IS plus specific equipment for control and measurement.

The architecture of a traditional industrial IS is shown in Figure 1.2, which is called Supervisory Control And Data Acquisition (SCADA) or Distributed Control System (DCS).

Components of an Industrial Control System 3

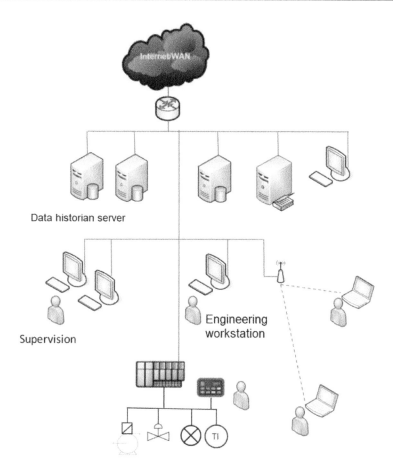

Figure 1.2. *Industrial information system. For a color version of this figure, see www.iste.co.uk/flaus/cybersecurity.zip*

The architecture of an industrial IS based on the Internet of Things (IoT) is shown in Figure 1.3. It is based on the Cloud and introduces new components and protocols described in the rest of the book.

In general, an ICS is composed of the same elements as a management IS with, in addition, specific equipment and specific software such as control command programs. These ensure real-time control and manage the archiving of data that characterize the evolution of the installation (history and alarm logs).

Henceforth, we will use the term ICS to refer to all the computer systems used to control a physical system.

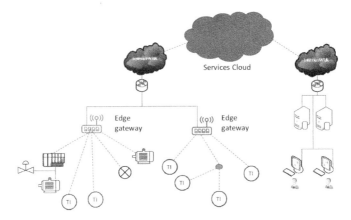

Figure 1.3. *IIoT information system. For a color version of this figure, see www.iste.co.uk/flaus/cybersecurity.zip*

1.1.4. Definition: IT and OT system

To distinguish ICSs from management ISs, the terms Information Technology (IT) and Operation Technology (OT) are often used.

An IT system is an IS as presented above. It is intended for the processing and storage of information. It is composed of computers, software, networks and human–machine interfaces (HMIs). An OT system is intended to interact with a physical system. It is composed of hardware and software to capture the evolution of a physical system via sensors, and to perform actions via actuators.

ICSs rely on varied information processing equipment, but integrate additional equipment to act on physical systems, and therefore belong to the OT world.

1.1.5. Definition: SCADA

A SCADA (Figure 1.4) is a system for controlling automation and data acquisition equipment. It makes it possible to centralize the control of an installation and to present an ergonomic HMI.

It includes workstations that generally run on Windows and are equipped with specialized software with a graphical interface that displays synoptics and trend curves. These workstations are connected via a network to equipment directly linked to the physical system: PLCs, input–output components, remote HMI, etc. It also includes servers, routers and other useful technical equipment. A SCADA is an ICS, but not all ICSs are SCADAs.

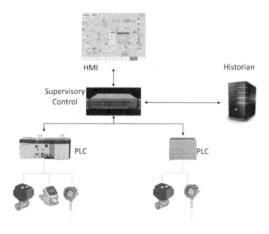

Figure 1.4. *The minimal functions of a SCADA. For a color version of this figure, see www.iste.co.uk/flaus/cybersecurity.zip*

1.1.6. *Definition: Distributed Control Systems (DCS)*

A DCS is a set of control systems connected in a network with a central unit to carry out supervision. The whole set is from the same manufacturer. Historically, this architecture was very different from the more heterogeneous SCADA systems, but with the evolution of technology, the differences have faded.

1.1.7. *Definition: Industrial Internet of Things (IIOT)*

The IoT refers to the extension of the Internet, that is, the global network for exchanging information with objects connected to the physical world. The Industrial Internet of Things (IIoT) is the use of the IoT in the industrial sector. These objects, more or less autonomous, will transform the industrial world. We sometimes talk about the plant of the future or Industry 4.0. The potential of this technology is enormous. The challenges posed by this interconnection are commensurate with the potential of this technology.

1.1.8. Different types of ICS

ICSs are used in many industrial sectors and critical infrastructures. A distinction is made between the manufacturing sector (e.g. the chemical industry or building management systems) and the distribution sector (e.g. of water or energy).

In the first case, the installation is geographically located. The manufacturing processes can be:

– continuous: these processes operate continuously, often with transitions to produce different qualities of a product, the quantities handled are real quantities. They are found in the chemical or petroleum industries;

– batch: these processes have distinct processing steps, performed on a given quantity of material. There is a separate start and end step for batch processing, with the possibility of short steady-state operations during the intermediate steps. Typical batch manufacturing processes include the manufacture of drugs or food;

– discrete: these systems generally perform a series of steps on a single device or a succession of machines to create the final product. The assembly of electronic and mechanical parts and the machining of parts are typical examples of this type of industry.

In the distribution sector, ICSs are used to control geographically dispersed assets, often over thousands of square kilometers, including water distribution systems, wastewater collection systems and energy management systems.

As explained above, an industrial IS is composed of elements similar to a traditional IS (workstations, servers, network equipment, printers, storage and backup systems) but also of specific elements designed to manage interaction with the physical system and provide an appropriate HMI: these include the PLC, remote terminal units (RTUs) and the acquisition system.

1.2. From the birth of the PLC to the SCADA system

Until the late 1960s, control and regulation of industrial manufacturing systems was carried out by mechanical or electromechanical relay-based systems. In 1968, Dick Morley, an engineer at GM, proposed the concept of a PLC, which he called Modicon (MOdular DIgital CONtroller). It is interesting to note that his motivation was to offer an alternative to minicomputers, whose prices were beginning to fall but which remained very complicated to program for the control of industrial systems. The cultural difference between the IT and OT world is therefore rooted in the very birth of the PLC.

The communication possibilities appeared in 1973, with the evolution of the 084 model into the 184 model. The first protocol developed was the Modbus protocol, from Modicon. Because of it, an automaton could communicate with other automatons and be very far away from the machine it was controlling. One of the first microprocessors was introduced in 1974 (Intel 8080) and, in 1975, Modicon introduced the 284, the first microprocessor-based device.

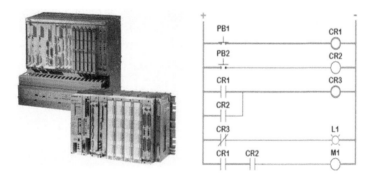

Figure 1.5. *Allen Bradley – Modicon 084. For a color version of this figure, see www.iste.co.uk/flaus/cybersecurity.zip*

The need to "supervise" the PLCs from a central station, based on a microcomputer, quickly became apparent. This resulted in an architecture composed of a PLC, directly connected to sensors and actuators running an "automation" program or performing PID control, and a system equipped with an HMI, which presents a synoptic of the system to be controlled and the trend of the evolution of the variables, archives data on mass storage and allows the program to be sent to the PLC. All the main functions of SCADA are present in this architecture.

ICSs then became more and more complex, which could be distributed over several sites. The solutions that have emerged have been classified into homogeneous solutions that are specific to a single manufacturer (i.e. DCS) or heterogeneous solutions consisting of PLCs, workstations and supervision software from different manufacturers (SCADA). Traditionally, homogeneous solutions are more commonly used in the process industry, whereas others are often used in the manufacturing industry. With the evolution of technology, the differences between the two types of solutions have narrowed and we talk about SCADA systems in a global way.

1.3. Programmable logic controller (PLC)

A PLC provides an essential function in a system's automation: it makes it possible to modify, trigger or modulate physical actions according to measured quantities, either by following a predetermined sequential operation, or by controlling or regulating quantities according to a fixed set point. These functions are those of the Basic Process Control System (BPCS). It operates in real time, in parallel with the evolution of the physical system. Loss of this ability to respond in real time to changes in the physical system is a problem, as there is a risk of unwanted evolution, and this represents a potential vulnerability.

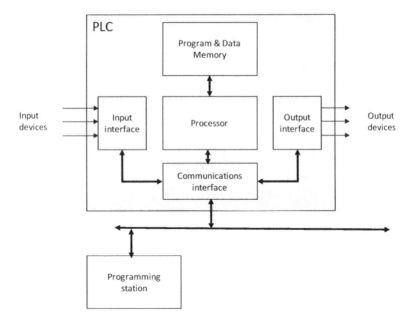

Figure 1.6. *Architecture of a PLC*

Generally, a PLC consists of a processing unit, a memory unit, a power supply unit, input/output interfaces and a communication interface. It can be connected to a programming device that is either a specific console or, in most cases, a PC workstation. As mentioned above, a PLC is a computer system with deliberately limited possibilities in order to make its implementation easier.

The architecture of a PLC is similar to that of a microcomputer without an HMI:

– the microprocessor-based central unit runs the program, reads the input signals and writes the output signals;

– the memory contains the PLC operating system, the system control program and the related data;

– the programming device is used to load the program into memory and access certain data areas;

– the input module makes it possible to receive signals from the system to be controlled. These can represent on-off values, for example, indicating that a switch is open or closed, or values that continuously vary within a range of values, for example, to describe a temperature. These signals can be received or transmitted as an electrical or computer signal via a field bus[1] using a specific protocol such as those described in Chapter 2 (EtherCat, CAN, Modbus, Profibus, Profinet, Ethernet/IP, etc.);

– the output module performs the reverse operation, and it transforms the outputs calculated by the PLC program into electrical or computer signals;

– the communication interface allows the PLC to communicate with the other PLCs, the supervision system (SCADA) and the programming station. Nowadays, it is often an Ethernet network interface.

From a functional point of view, a PLC is the basic equipment for the control and regulation of physical systems. There are two main types of features:

– the regulation of continuous systems;

– the automation of sequential systems.

In the first case, the objective is to bring a physical quantity to evolve according to a desired (possibly constant) profile. The system measures the

---

1 This bus is a computer bus that can potentially be attacked with computer means when connected to the higher level industrial network. The electrical connections would require an electrical device to read or modify the values and are less vulnerable.

controlled quantity in a relatively small period of time, in an almost continuous way, and calculates one or more values of the actions to be applied to the physical system to bring the controlled value to the desired value. This is the case, for example, of a temperature regulation that can be modulated by opening a heat transfer fluid valve to obtain a desired temperature.

In the second case, the PLC performs actions from on–off inputs. It can either perform actions according to measured quantities, for example shutting off a valve, if a container is filled or perform action sequences to control a device depending on the measurements and a manufacturing recipe.

In all cases, a PLC operates on a regular cycle (in the order of 10–100 ms), during which it reads the values of the inputs it copies to the input memory (IM), then executes the program, puts the results into the output memory (OM) and sends these values to the physical device (Figure 1.7). The image of the state of the physical device is therefore stored in the PLC's memory. This cycle also includes various operations to ensure the proper functioning of the system: integrity test of the I/O module, verification that the user program is not modified, verification that the system is not blocked via a watchdog and communication operations via the interface to the remote modules, the programming station and the HMI interfaces.

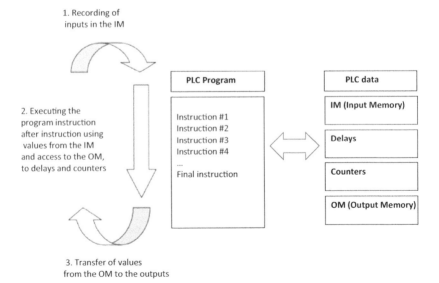

Figure 1.7. *Execution cycle*

Components of an Industrial Control System   11

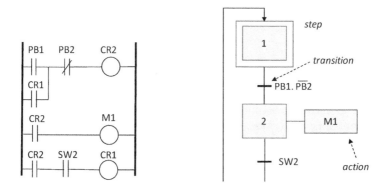

Figure 1.8. *Example of LD and SFC language*

To simplify the use of PLCs and to remain as close as possible to the electromechanical systems mentioned in the introduction, a number of specific languages have been defined. They are described in IEC 61131/3. They are as follows:

– The ladder diagram (LD) (Figure 1.8);

– The functional block diagram (FBD);

– The sequential function chart (SFC) (Figure 1.8) ;

– The structured text (ST), close to the PASCAL language;

– The instruction list (IL), close to the assembler.

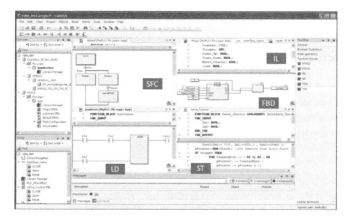

Figure 1.9. *Programming languages IEC 61131/3. For a color version of this figure, see www.iste.co.uk/flaus/cybersecurity.zip*

Malware can be written directly in these languages (Govil *et al.* 2018). The loading of the program into the PLC is not very secure, so this is a major vulnerability.

These languages can either be interpreted by the PLC or compiled on the development station and uploaded to the PLC. When considering cybersecurity, it should be taken into account that a lower level computer system is required to run them on the PLC. Most of those available on the market, whether PLCs, RTUs, safety instrumented systems (SISs), or DCSs, now have a commercial operating system. Here, for example, are some used by major manufacturers:

– Schneider Quantum: VxWorks;

– Siemens: VxWorks from 2014;

– Allen-Bradley: PLC5: Microware OS-9, Controllogix: VxWorks;

– Emerson DeltaV: VxWorks.

The vulnerabilities of these operating systems are therefore vulnerabilities for the equipment that uses them, and several manufacturers can be affected by the same malware. An example is the ICSA-15-169-01 (TCP Predictability Vulnerability) vulnerability that affected VXWorks. Such a system is used on more than 100,000 devices connected to the Internet[2], as well as on systems as diverse as network routers, the Boeing 787 Dreamliner or the Gran Telescopio Canarias telescope.

1.4. RTU, master terminal unit and intelligent electronic device

A Remote Telemetry Unit (RTU) is a microprocessor-controlled electronic device that connects a physical system to a master system, typically a PLC, master terminal unit or SCADA. It transmits telemetry data and receives messages to control the physical system. Communication is often provided via a modem, cellular connection, radio or using communication technology for long distances. Often, the available electrical energy is limited, for example energy from solar panels, and this is one of the constraints to be respected for these systems.

An RTU may also be called a remote control unit. From a cybersecurity perspective, these systems have vulnerabilities similar to PLCs, are sensitive to vulnerabilities on the link and are physically vulnerable because they are isolated.

---

2 Available at: www.shodan.io.

In the field of electrical power distribution, the remote control units used to control the distribution are called intelligent electronic devices (IED). For example, they act on circuit breakers or transformers.

1.5. Programmable Automation Controller

The term Programmable Automation Controller (PAC) was introduced in the early 2000s. A PAC is a PLC with extended capabilities. It offers more processing possibilities: a greater number of PID control loops, advanced control possibilities (fuzzy controller, predictive control, etc.), batch control modules and even dedicated business modules. It also has more connectivity and can manage remote modules similar to RTUs.

With technological developments, PLCs have become more and more sophisticated and the boundary between PLCs and PACs is blurred. This term is often used in IIoT architectures.

1.6. Industrial PC

An industrial PC is a classic PC reinforced from a hardware point of view and often supplied in a chassis adapted to the industrial world (rackable and fanless, for example). It often has a larger number of inputs/outputs (I/Os), and some of them have specific possibilities, such as analog ports. It can use a traditional operating system, a dedicated system such as Windows Embedded or a real-time operating system (VxWork or FreeRTOS, for example).

1.7. Safety instrumented systems

An SIS (Goble and Cheddie 2005) is defined as a system composed of sensors, logic processing and actuators designed for:

– ensuring that an industrial process evolves automatically to a safe state when specified conditions are violated;

– allowing a process to evolve safely when the specified conditions allow (permissive functions);

– take measures to mitigate the consequences of an industrial hazard.

Given the importance of an SIS to the safety of a facility, it is designed so that its probability of failure is less than a value based on the severity of the impact of a loss of control. This approach, carried out as part of the functional safety study, is described in Chapter 8.

From a technological point of view, SISs are often built using specific PLCs, with reinforced and redundant hardware. One example is the Triconex[3] brand, which markets such devices. The functions performed by the SIS are critical to the security of a facility. If the BCPS fails to maintain the system in a safe area, the SIS must take appropriate action. It corresponds to a level of defense-in-depth protection of the physical system (Figure 1.11).

SISs should, therefore, be considered with particular attention to cybersecurity. As these systems have the role of bringing the physical system back to a stable, safe state, they represent a particular challenge for a cyber attacker. A simple denial of service attack can significantly increase the level of risk of the system, since in the event of a BCPS malfunction, the security actions normally triggered by the SIS can be prevented and significant damages can occur.

It should be noted that SISs can also be the targets of attacks in order to create unintentional shutdowns of the production system. Indeed, as the SIS takes control of the normal control system, it is possible, by attacking it, to override the actions of the normal control system and cause the system to shut down.

It therefore appears that protecting the SIS is a priority in an ICS.

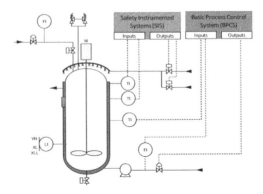

Figure 1.10. *Safety instrumented system (SIS). For a color version of this figure, see www.iste.co.uk/flaus/cybersecurity.zip*

---

3 Brand marketed by Schneider.

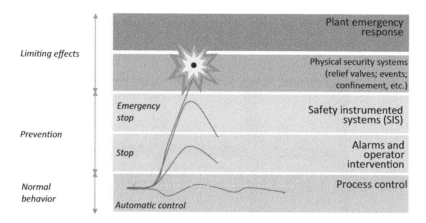

Figure 1.11. *The position of the SIS in terms of protection level. For a color version of this figure, see www.iste.co.uk/flaus/cybersecurity.zip*

1.8. Human–machine interface (HMI)

HMI is an essential part of ICSs. It allows the user to visualize how the system works and take the necessary actions. Once constructed physically, with entire walls covered with indicators, dials and adjustment buttons, it has been replaced by graphic screens, when technological developments have made it possible. Beyond ergonomics issues, the software and stations used by these HMIs are also a source of vulnerabilities. The HMI makes it possible to monitor the system by observing the different quantities and their evolution, as well as to control the system by launching sequences or modifying the setpoints. It is possible via these interfaces to act maliciously on the controlled system, and managing the access control to these stations, either physically or via a remote connection, is an important point for cybersecurity.

There are several types of HMI: the first are PC workstations running on a traditional operating system such as Windows and using software to present synoptics and trend curves. These software are often used in SCADA supervision stations. The vulnerabilities of these HMIs are those of traditional computer systems.

The second type of HMI is made up of dedicated units running on an embedded operating system such as Windows Embedded and, often, with a touch screen. They are often close to the physical system and can be more or less physically isolated. Cybersecurity issues are those related to embedded systems, one of the problems being to carry out updates in a systematic and secure way.

Figure 1.12. *HMI on PC. For a color version of this figure, see www.iste.co.uk/flaus/cybersecurity.zip*

HMIs interact directly with PLCs and servers of the ICS. The link and the protocol used can be a source of vulnerabilities.

Figure 1.13. *Dedicated unit for HMI. For a color version of this figure, see www.iste.co.uk/flaus/cybersecurity.zip*

1.9. Historians

In control systems for centralized (DCS) or heterogeneous (SCADA) industrial systems, it is useful to store a certain amount of information such as the evolution of the different values that have been collected from the sensors, the evolution of setpoints, alarms, the status of equipment, etc. These data are stored in specific databases, optimized for time series. Such software is marketed by companies offering complete ICS solutions such as ABB, Honeywell, Schneider, Siemens or by specialized companies such as Panorama[4] or AspenTech[5].

From a cybersecurity perspective, it should be noted that access to these databases is provided from both the OT and IT systems, and that this aspect must be taken into account, as it effectively introduces a connection between the two types of network.

1.10. Programming and parameter setting stations

The physical system is controlled by the PLC in real time. The program to be executed is developed in a programming environment dedicated to the brand of the PLC. For example, for Siemens S7 PLCs, there is the STEP 7 environment running on a PC. For Schneider PLCs, we find Unity Pro for the Modicon range, Concept for Quamtum, TwidoSuite for Twido.

These software programs make it possible to write the program in at least one IEC 1131-3 language, load or read this program to and from the PLC, read the PLC memory, perform various basic operations such as stopping or restarting, setting the time, and configure controller and automation parameters such as PID constants and timers (Flaus 1994).

These development stations and communication with PLCs are therefore interesting gateways for attackers. Malicious software capable of using the development system or intercepting communication with the PLC can very easily take control of it[6]. These workstations often run on Windows and have the vulnerabilities found on traditional computer workstations.

---

4 Available at: https://uk.codra.net/panorama/.
5 Available at: http://home.aspentech.com/.
6 This is how Stuxnet was implemented in the PLC.

1.11. Industrial Internet of Things (IIoT)

The notion of the IoT (Minerva *et al.* 2015) is similar to that of CPS. It is defined by the NIST (Cyber-Physical Systems 2017) as an intelligent system that includes networks of physical and computer components that interact with one another.

IIoT devices are the end-points for the Internet of Things, those that are connected to the physical world. They are characterized by:

– their acquisition and control capabilities, which ensure interaction with the physical world. Different types of sensors can measure a number of quantities (temperature, pressure, position, etc.) and are coupled to an analog-to-digital converter for acquisition. In the other direction, a digital-to-analog converter is used to control actuators (switches, motors, valves, etc.) or different output devices (LED, display, loudspeaker, etc.);

– their processing and storage capabilities: a processor, which can be relatively powerful, is coupled with RAM and non-volatile flash memory;

– their connectivity: an IoT device has features to connect to a network, either traditional (wired or Wi-Fi) or more specific like LPWAN (Chapter 2), for example;

– their energy management: not all devices are connected to a power source and some operate on batteries, such as isolated sensors, or some HMI devices such as switches. Power can also come from ambient energy trapping and be limited;

– their ability to be physically secured through a secure storage device, called a Secure Element. It is a tamper-proof hardware element, capable of securely hosting applications and storing confidential and cryptographic data;

– their encryption capabilities, which depend on the processor's computing capabilities, the presence or not of a dedicated circuit, and of the energy management.

Many ready-to-use devices called Component Off-The-Shelf (COTS), including microcontrollers and single board computers, are designed around integrated circuits called System-on-a-Chip (SoC), which include most of the features presented above. The available computing power is relatively high, with a frequency of about 100 MHz to several GHz, a flash memory of 32 MB to 1 MB and a RAM of at least 128 KB.

The equipment must be uniquely identifiable. Depending on the type of network used, identification systems can be:

– unique identifiers stored in the device at the time of manufacture;

– computer identifiers such as IP addresses or MAC addresses;

– identifiers from the world of telephony: SIM card identifiers and mobile numbers;

– Radio-Frequency Identification (RFID) or Near-Field Communication (NFC) identifiers.

These devices are often isolated and their physical vulnerability is significant. They are, therefore, exposed to attacks based on reverse engineering or physical attacks, such as auxiliary channel attacks on consumption, as described in Chapter 4.

1.12. Network equipment

The transfer of information between the different components of the ICS is ensured by network equipment, wired or wireless, to which the various equipment is connected.

It is important to note that these devices work with operating systems, either proprietary or derived from Linux (DD-WRT, for example) and are as vulnerable as other devices.

1.12.1. *Switch and hub*

Switches are the foundation of most corporate networks. Switches allow the different connected devices to communicate with each other. They manage the data flow on a network by transmitting a received network packet only to the devices for which the packet is intended. Each network device that is connected to a switch can be identified by its network address, allowing the switch to direct the flow of traffic while maximizing network security and efficiency.

There are remotely manageable switches, which is an additional potential vulnerability for these devices. Some are able to copy streams and transmit them to an intrusion detection system.

1.12.2. *Router and gateway*

A router is a network device that transfers data packets between computer networks. Routers perform Internet traffic management functions. A data packet is usually transmitted from one router to another over networks, which constitute an inter-network (network of networks), until it reaches its destination node.

A router is connected to two or more data lines from different networks. When a data packet arrives on one of the lines, the router reads the network address information in the packet to determine the final destination. Then, using the information in its routing table, it directs the packet to the next router or to the destination equipment. A router acts as a dispatcher. It analyzes the data sent through a network, chooses the best route for the data to be transported and sends it.

A gateway connects a network with one or more other networks and can convert protocols if necessary. The most common task of a gateway is to be what is called the "default gateway", the router to which all packets are sent, when there is no other local route that can be associated with them.

Nowadays, a gateway manages the connection of a Local Area Network (LAN) with the Internet and is therefore similar to a router. A few decades ago, a gateway was responsible for translation between different types of networks such as Ethernet and Token-Ring.

1.12.3. *Firewall*

A firewall is a device that filters incoming and outgoing traffic on a network at the level of a gateway, a router or a specific equipment. It is based on a packet filter that operates on layers 3 and 4 of the OSI model (Chapter 2, section 2.2.4), and decides which packets should pass, be rejected or redirected, based on a set of predefined rules.

A firewall is often implemented as a specific software module in a gateway or router running on Linux. Industrial firewalls can be implemented with dedicated equipment.

1.12.4. *IoT gateway*

An IoT gateway is a physical device or software that serves as a connection point between the Cloud and the IIoT equipment: aggregators, sensors and PAC. All data transferred to or from the Cloud pass through the gateway, which can be a dedicated hardware device. An IoT gateway can also be called an "intelligent gateway".

It may have some computing power and perform initial processing on the data before sending it to the cloud (*fog computing*, Chapter 2). This minimizes the amount of data being reported.

Another advantage of an IoT gateway is that it provides additional security for the IoT network and the data it carries. Since the gateway manages information moving in both directions, it can protect data transferred to the cloud against a lack of confidentiality by encrypting it, and protect IoT devices against malicious external attacks by filtering flows and providing intrusion detection capabilities.

Finally, these gateways can validate the rights of IoT devices when they are added to the network (provisioning).

1.13. Data processing platform

IoT gateways are connected to platforms that receive, store and use data. Many platforms, offered by generalist operators and specialists in a field (maintenance for example), are available in the cloud.

These platforms enable the management of data transfers, the use of analysis software, storage management and the provision of services that exploit these data.

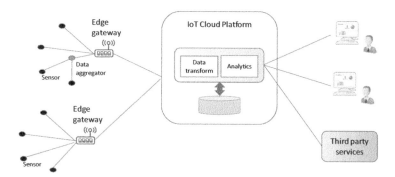

Figure 1.14. *IoT platform. For a color version of this figure, see www.iste.co.uk/flaus/cybersecurity.zip*

1.14. Lifecycle of an ICS

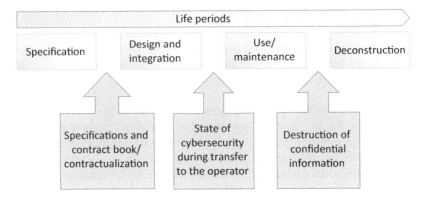

Figure 1.15. *Life phase*

Taking into account the lifecycle of the ICS is important for the control of cybersecurity. The lifecycle of installations is quite long, which means that old equipment has to be managed, which can be a security issue. In addition, significant changes may occur on the ICS during major changes in the manufacturing process, which may have consequences for cybersecurity. Therefore, it is necessary to manage security throughout the lifecycle of the ICS (PA Consulting Group 2015). In a rather classic way, we distinguish four key periods: specification, design and integration; operation with associated maintenance and, finally, decommissioning.

During the specification phase, it will be necessary to specify the requirements expected in the specifications. Most of the time, the systems are designed by external service providers. In this case, the security requirements must be explicit and included in the specifications and contracts.

The design and integration phase must take into account security and in particular:

– the identification of critical assets;

– the definition of a secure architecture;

– the definition of zones and conduits;

– the choice of equipment with sufficient security capacities;

– the definition of basic measures to secure equipment;

– the possibility of carrying out preventive and curative maintenance operations in safe conditions at an adequate level;

– taking into account physical security (via the location of critical equipment);

– the definition of roles for stakeholders.

A penetration audit and test program is recommended at the end of this phase.

Following these two phases, the system is transferred to the operator during commissioning. During this step, it is recommended to carry out an exhaustive inventory of the system's cybersecurity level and to ensure that the available means are available to maintain it at an acceptable level (ANSSI 2013a). For critical systems, approval is required.

The ICS then moves into the operational phase in which cybersecurity must be controlled, following an approach such as those presented in Chapters 8 and 11.

The decomissioning phase, particularly when it is partial, is also very important for cybersecurity. Indeed, during this phase, confidential information may be made available to malicious persons, for example, staff names and addresses, passwords, technical specifications or configuration information and customer data.

2

Architecture and Communication in an Industrial Control System

2.1. Network architecture

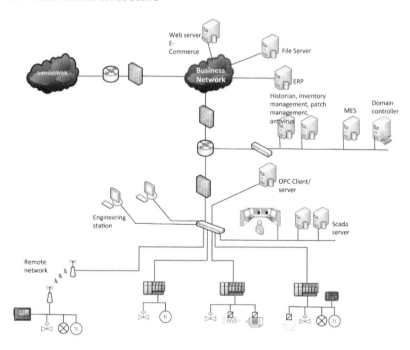

Figure 2.1. *Typical ICS architecture. For a color version of this figure, see www.iste.co.uk/flaus/cybersecurity.zip*

2.1.1. *Purdue model and CIM model*

The typical architecture (Figure 2.1) of an industrial control system includes several types of networks. We can distinguish between

– the field network that connects sensors and actuators to programmable logic controllers (PLCs);

– the control network that connects the PLCs and associated equipment, such as the human–machine interface (HMI) and the supervision system;

– the production network that links the various control networks of the site and the manufacturing execution systems (MES) or historical servers;

– the corporate network (or IT network).

A number of models have been proposed to structure this architecture and organize it hierarchically. These models simplify reality a little and introduce a level of decomposition that is not always so clear, but they are useful and serve as a basis for the breakdown into zones.

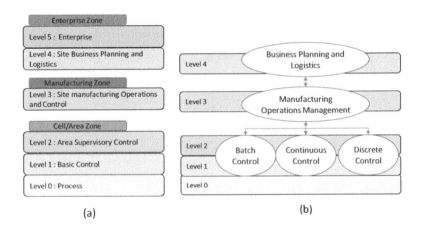

Figure 2.2. *(a) Purdue and (b) ISA85 models. For a color version of this figure, see www.iste.co.uk/flaus/cybersecurity.zip*

The first of these models is Purdue's model (Williams 1994). It introduces five levels:

– level 0 (physical process): this level corresponds to the physical systems used for production. The sensors and actuators are located at this level;

– level 1 (local or basic control): this level includes the functions involved in the detection, observation and control of the physical process, which are performed by information processing systems such as PLCs, Remote Terminal Units (RTUs), etc. The latter read the data from the sensors, execute algorithms if necessary and store the state of the physical system. They can perform continuous control, discrete control, sequential control or batch control. The IEC 62443 model (Figure 2.3) explicitly shows the security (SIS) and protection systems in this level 1. The SIS are responsible for monitoring the process and automatically returning it to a safe state if it exceeds security limits. They also have the function of alerting the operator in the event of imminent hazardous conditions;

– level 2 (supervision control): this level includes the functions involved in monitoring and controlling the physical process. It includes the HMIs of control and data acquisition systems (SCADA) and distributed systems (DCS). Equipment at this level is generally associated with the production area;

– level 3 (operations management): this level includes the functions involved in flow management to achieve the desired production. These include batch management systems or MES, as well as historical databases or site-wide optimization or quality management systems. Part of the supervision system may also be at this level (Figure 2.3);

– level 4 (enterprise business systems): this level includes the functions involved in the management of manufacturing and processing operations. Enterprise resource planning (ERP) is the main system used at this level. It manages basic production planning, raw material use, transport and stock levels. In general, the ERP provides the MES with a list of manufacturing orders that the MES will have the task of executing, controlling the flow of operations and their traceability, providing the ERP with material consumption, production performance and the quality of the manufactured products.

Purdue's model provides a reference model for the hierarchical representation of the company's control systems. It is easily adaptable by end users, system integrators and OEM suppliers to integrate their offerings into the enterprise levels. Such adaptation (Figure 2.4) is proposed, for example, by the alliance between Cisco and Rockwell Automation, which have integrated Purdue's model into their Converged Ethernet or Converged Plantwide Ethernet solution architectures. Compared to the initial version, a demilitarized zone (DMZ) appears between levels 3 and 4 to separate the company's industrial (Operation Technology [OT]) and IT networks.

This architecture is often used as a basis for the implementation of a secure architecture (Chapter 10).

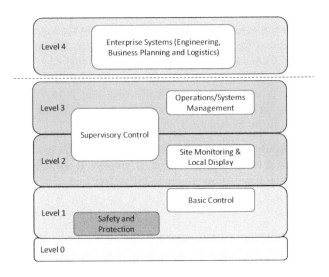

Figure 2.3. *IEC 62443 model. For a color version of this figure, see www.iste.co.uk/flaus/cybersecurity.zip*

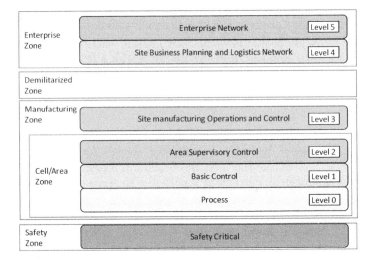

Figure 2.4. *Converged Plantwide Ethernet (CPwE) model. For a color version of this figure, see www.iste.co.uk/flaus/cybersecurity.zip*

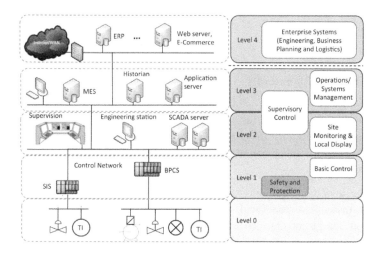

Figure 2.5. *Model of the previous installation according to the CIM architecture. For a color version of this figure, see www.iste.co.uk/flaus/cybersecurity.zip*

2.1.2. Architecture of the Industrial Internet of Things

The Internet of Things (IoT) is a system composed of computer devices connected to each other. Each object has a unique identification and the ability to transfer data over a network without requiring interaction between humans or between humans and machines.

In an industrial context, objects are as close as possible to the physical world and allow data to be acquired or actuators to be controlled. These elementary objects can exchange data with equipment called "aggregators" or gateways that act as relays to transfer data via the Internet to processing and storage platforms. These data and services are made available to the various users.

The objects are therefore not directly connected to the Internet, but form subnetworks that are themselves connected to the Internet. The protocol used by these subnetworks is often specific, more adapted to the IoT in terms of distance or energy needs (section 2.2) than the Transmission Control Protocol (TCP)/Internet protocol (IP) stack used by the Internet (Figure 2.9).

The tendency of Industrial Internet of Things (IIoT) is to separate low-level tasks (measurements, actions) and algorithmic aspects that are implemented on a platform accessible via the network. Most of the IIoT equipment are therefore sensors, actuators or Programmable Automation Controller (PAC).

The architecture of an IIoT system (Lin *et al.* 2017; Voas 2016) (Figure 2.6) is formed of three levels:

– the level that groups objects, aggregators and distribution centers. At this level, different types of networks can be used to meet field constraints (energy consumption, distance) and to ensure communication between objects and equipment connected to the traditional IP data network, called edge devices;

– the level corresponding to the processing and storage platform. At this level, transfers use traditional computer networks, and storage is done in the cloud on storage and data processing platforms;

– the level corresponding to use of the data: this can be done by different components of the company, in particular by OT users for the control and maintenance of the installations.

It can be noted that the IIoT architecture is not a simple direct connection of devices to the Internet, but is carried out via a set of gateways that have more or less important data processing capabilities. These gateways also allow the various protocols used by IIoT (section 2.2) to be converted to IP.

Data processing can be carried out at several levels:

– Cloud computing consists of processing data at the storage platform level;

– Fog computing is processing performed on the gateways of the local network (IoT gateway). It is essentially an intermediate layer between the cloud and the hardware to allow initial processing in order to reduce the amount of data to be transported to the cloud;

– Edge computing is carried out at the level of devices in contact with the physical system, for example, in PAC in order to guarantee a certain response time and independence from communication malfunctions.

The choice of the processing level is made depending on the necessary reactivity, the amount of data to be transmitted and whether or not the local data are sufficient to carry out the treatment.

These different levels of treatment can be illustrated in the context of an intelligent lighting system, which operates on the basis of movement. When there is motion detected, data are sent and must be processed to turn the lamps on or off. It is more appropriate to perform this processing locally (at the periphery, edge computing). Switching on or off according to brightness is more global, and it is preferable to process it at the gateway (fog computing). The company that operates the intelligent lighting system may also want to monitor the energy efficiency and lifetime of the lamps. The data that provide this overview of intelligent lighting usage are sent to the cloud to generate usage reports.

This type of infrastructure with local processing (edge or fog computing) is well suited for industrial control systems in which reflex functions can be performed locally, while more advanced processing will be performed in the cloud.

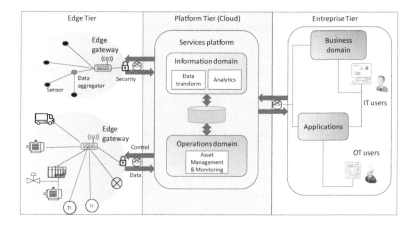

Figure 2.6. *IIoT architecture. For a color version of this figure, see www.iste.co.uk/flaus/cybersecurity.zip*

2.2. Different types of communication networks

2.2.1. *Topology*

The exchange of information between different computer equipments is done via a physical wired connection, a wireless connection or exchanges of mass memory (USB sticks, disk, etc.). In the first two cases, we speak of a computer network[1].

---

1 In the third case, we sometimes speak of a "sneaky network".

The first step in building a network is to define how the workstations are connected to each other. There are different topologies: star, bus, mesh (partial or total), ring. Depending on the topology chosen, it will be easier to wire or more robust against connection failures.

The most common topology encountered for workstation networks is star topology, which requires a concentrator called a "switch". Bus topology is found when equipment is connected from one person to another, which is the case with sensor or actuator networks. Ring topology is used for industrial networks, for which a "self-healing" function offers a certain redundancy, and mesh topology is used to obtain a better robustness.

A mixed topology presented in Figure 2.8 is found in IIoT architectures. Each IIoT device subnetwork is connected to a gateway with a star or mesh architecture, and the gateways are connected in a star configuration to a higher network or to the Internet to provide the connection to the cloud.

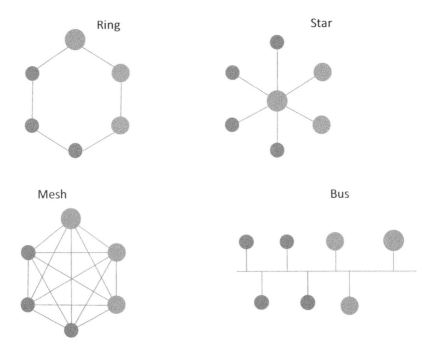

Figure 2.7. *Classic network topologies. For a color version of this figure, see www.iste.co.uk/flaus/cybersecurity.zip*

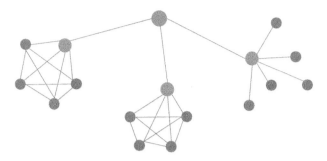

Figure 2.8. *Mixed network topology (IIoT). For a color version of this figure, see www.iste.co.uk/flaus/cybersecurity.zip*

2.2.2. *Types of networks*

There are different types of networks that are classified according to their scope and maximum throughput:

– a Local Area Network (LAN) is a corporate network that is usually limited to a building, a floor or a room. In modern networks, most computers are connected to a local network via one or more switches. Several local networks can be connected via a router or a virtual private network (VPN) (section 2.2.3). A LAN can be wired or wireless and, in this case, it most often uses the Wi-Fi protocol;

– a Personal Area Network (PAN) is a network that covers a small area, of the order of a few tens of meters, centered around a person and uses protocols such as Bluetooth classic or Bluetooth Low Energy (BLE), IEEE 802.15.4, Zigbee or ISA100.11a, which are detailed below;

– a Low-Power Wide Area Network (LPWAN) is a network that can have a range of up to 10 km and uses protocols such as LoRaWAN or SigFox. The flow rate is limited, but it uses little energy.

We also find the following acronyms:

– a Body Area Network (BAN) is a network of devices attached to the body, such as medical devices or gadgets;

– a Wide Area Network (WAN) is a network extended over several countries or even worldwide, such as the Internet;

– a Metropolitan Area Network (MAN) is a network deployed at the scale of an urban area, often consisting of several LANs.

The networks used in ICSs are essentially LAN networks, wired or wireless. PAN networks are used to connect terminal equipment such as sensors to a gateway connected to the LAN network.

In IIoT installations, there are also LPWAN-type networks, combining devices such as sensors, with low-energy resources, but requiring only a reduced data transfer rate.

2.2.3. Virtual private network

A VPN is a network that relies on an existing public infrastructure, such as the Internet, to which it adds a set of security mechanisms based on encryption and/or authentication. Almost all VPNs provide the capability to secure access to an entire network and, because of powerful cryptology, they also protect against espionage and manipulation. They work on the TCP/IP stack (Figure 2.9) on layers 3, 4 or 7. Appendix 1 provides more details on how encryption security works.

Typical protocols or protocol stacks are IPsec, PPTP and OpenVPN. Generally, they are used to connect external agencies and to integrate mobile employees. A VPN can be considered secure, but the source or destination may not be, and they may be contaminated by malicious software. As the communication is encrypted, any possible detection of suspicious exchanges is impossible. The use of a VPN does not therefore solve all problems.

2.2.4. OSI model

The Open Systems Interconnection (OSI) model is a conceptual model for standardizing the internal functions of a communication system by partitioning it into layers of abstraction. It defines seven layers:

– *physical*: the physical layer ensures transmission through a physical, electrical, wired, optical or radio transmission means;

– *data link*: the data link layer provides data transfer from node to node (between two directly connected nodes), and also manages the error correction of the physical layer. It manages MAC addresses (which are identifiers linked to a machine or a network interface of a machine);

– *network*: this layer is responsible for the transmission of packets, including routing through different routers. It manages the logical addressing mechanism (IP address);

– *transport*: the transport layer manages the coordination of data transfer between hosts using a protocol such as the Transport Control Protocol (TCP) or the Unified Datagram Protocol (UDP). It manages the retransmission in case of error;

– *session*: when two devices (computers or servers) need to exchange, a session must be created in order to keep information on the status of the current transmission, and this is done at the session layer. The functions of this layer involve configuration, coordination (how long a system must wait for a response, for example) and termination;

– *presentation*: the presentation layer prepares the data and, in general, ensures translation of the application format into the network format. In other words, the layer "presents" the data for the application or network. A good example of presentation is the encryption and decryption of data for secure transmission, which occurs at layer 6;

– *application*: this layer provides services (http, ftp, smtp, etc.) for applications with which users interact directly.

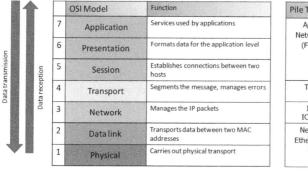

Figure 2.9. *OSI models and TCP/IP stack. For a color version of this figure, see www.iste.co.uk/flaus/cybersecurity.zip*

2.3. Transport networks

2.3.1. *Ethernet*

Over the years, Ethernet has become the standard protocol for wired communication. It operates at level 2 and uses the MAC addresses of the various connected devices to transport the data. The IEEE 802.3 standard defines the standard version of Ethernet with the carrier sense multiple access/collision detection (CSMA/CD) protocol for sharing physical access to the network. This protocol is not real time in the sense that the time for a transmission cannot be guaranteed in a deterministic way.

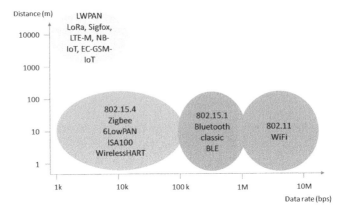

Figure 2.10. *Different wireless communication solutions. For a color version of this figure, see www.iste.co.uk/flaus/cybersecurity.zip*

2.3.2. *Wi-Fi*

Wi-Fi is a set of communication protocols defined by the IEEE 802.11 standard, allowing a wireless local area network to be built. Physical transport is provided by radio waves. Wi-Fi operates at level 2 of the OSI model, and is often coupled to the TCP/IP.

Wi-Fi makes it possible to connect computers and communicating objects or even peripherals to a broadband connection: from 11 theoretical Mbit/s or 6 real Mbit/s in 802.11b, to 54 theoretical Mbit/s or about 25 real Mbit/s in 802.11a or 802.11g, and even up to 1.3 theoretical Gbit/s for the 802.11ac standardized in 2013.

The range can reach several tens of meters indoors, if there are no obstacles.

2.3.3. *The IEEE 802.15.1 (Bluetooth) standard*

Bluetooth is a communication system born in the 1990s, operating in the free 2.4 GHz band and defined by the IEEE 802.15.1 standard, and is a robust solution that allows devices to be connected in master-slave mode (maximum seven slaves), over a range of a few meters to a few tens of meters, with a maximum throughput of 2.1 Mbit/s in the V2.1 version. The V3.0 version, called "high speed", based on an 802.11 Wi-Fi layer can reach up to 24 Mbit/s. This version supports IP frames.

A low-power version called BLE has been developed for the IoT. Version 5, adopted in 2016, increases the range to 300 m, with a throughput of 2 Mbit/s.

2.3.4. IEEE 802.15.4 networks

802.15.4 is a communication protocol defined by the IEEE for connected objects. It is intended for wireless networks of the Low Rate Wireless Personal Area Network (LR WPAN) family. These are short-range, low-speed networks for low-power devices. The range is about a few tens of meters with a throughput of 250 kbit/s. It offers a solution for the lower layers that is based on the Zigbee solution.

The 802.15.4 standard is used by many implementations based on proprietary protocols or IP, such as the Zigbee protocol and several industrial solutions:

– WirelessHart, proposed in 2007 by a consortium called Hart Communications Foundation (HCF) with more than 210 members;

– ISA100.11a, proposed by the ISA in 2009 and approved by a committee representing 250 companies.

These solutions add mechanisms that provide better robustness, availability, and real-time performance adapted to the requirements of industrial applications. These mechanisms are mesh network and deterministic network mechanisms. The ISA100.11a standard ensures compatibility with 6LowPAN, which is a standard defining an adaptation layer between IP6 and 802.15.4 networks (see section 2.6.1).

These protocols are used as protocols for IIoT field networks between objects and concentrators.

2.3.5. *LPWAN networks*

The LPWAN family of networks is designed to connect devices that require little bandwidth and have low-energy resources. In this family, the most well-known solutions are the LoRa and Sigfox systems.

The range can be up to 10 km, and the flow rate is quite low. For example, it is around 100 bps for Sigfox and 0.25-50 kbps for LoRa. The devices targeted are low-energy sensor networks.

2.3.6. *Cellular networks*

In the meantime, a number of cellular network standards have been introduced for LPWAN networks. These are as follows:

– the LTE-M (Long-Term Evolution) standard, also known as eMTC (Enhanced Machine Type Communication), a low-speed version that uses less energy than the LTE standard (marketed as 4G);

– the NB-IoT standard, with a maximum throughput of 60 kbps and a positioning close to LoRa and Sigfox;

– the EC-GSM-IoT standard, which is based on the GSM (2G) standard and supports data rates between 350 bps and 70 kpbs.

The 5G has been designed to deliver many improvements: a significant reduction in energy consumption, low latency (less than millisecond), and throughput up to 10 Gbps.

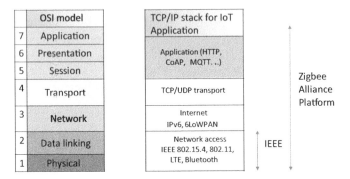

Figure 2.11. *IIoT protocols and OSI model. For a color version of this figure, see www.iste.co.uk/flaus/cybersecurity.zip*

2.4. Internet protocols

2.4.1. *The Internet protocol*

IP is one of the main protocols used by the Internet. The most widely used version is the IPv4 version. Like Ethernet, it is a stateless protocol (no session information), which means that it does not know any relationship between packets. This protocol is used to define the source and destination host on layer 3, to find the (fastest) path between two communication partners, to route packets and to handle errors with Internet Control Message Protocol (ICMP). For example, an error occurs if the destination is inaccessible.

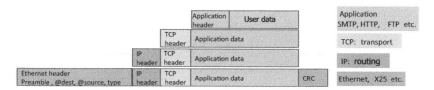

Figure 2.12. *Data packages. For a color version of this figure, see www.iste.co.uk/flaus/cybersecurity.zip*

An IP frame contains information about the source, destination, technical information and the data itself (Figure 2.13). An IPv4 address is defined by four bytes (e.g. 192.168.1.1.2).

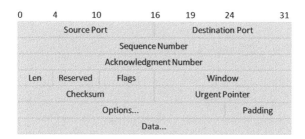

Figure 2.13. *IP frame*

2.4.2. *Transmission Control Protocol*

TCP layer provides session management. A new TCP session is initiated by a "three-way handshake", which offers a well-known vulnerability. TCP numbers all packets to ensure that they are processed in the same order as the one transmitted by the source system. The destination host sends an acknowledgment to inform the source that the packet has been received correctly after checking a checksum, otherwise the source retransmits the packet. Finally, TCP allows the use of ports. The port of the sending instance is called a "source port" and the receiving port on the target is called a "destination port". Commonly used application protocols such as HTTP, FTP, IRC, etc., have a default port of less than 1024. For example, an HTTP server normally listens on port 80.

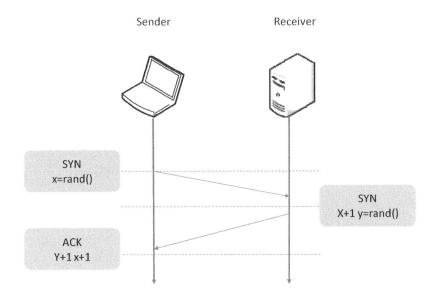

Figure 2.14. *Three-way handshake*

Operation of a "three-way handshake" is as follows:

– SYN: the client that wants to establish a connection with a server will send a first SYN (synchronized) packet to the server. The sequence number of this packet is a random number x;

– SYN-ACK: the server will respond to the client using a SYN–ACK (Synchronize–Acknowledge) packet. The ACK number is equal to the sequence number of the previous packet (SYN), incremented by one (x + 1), while the sequence number of the SYN-ACK packet is a random number y;

– ACK: finally, the client will send an ACK packet to the server, which will serve as an acknowledgment of receipt. The sequence number of this packet is defined according to the value of the acknowledgment received previously (for example: $x + 1$) and the ACK number is equal to the sequence number of the previous packet (SYN-ACK), incremented by one ($y + 1$).

A SYN flood attack works by not responding to the server with the expected ACK code. The malicious client either cannot send the expected acknowledgment of receipt or, by usurping the source IP address in the SYN, force the server to send the SYN-ACK to a fake IP address, which will therefore not send an ACK, because it has never sent a SYN. The server will wait for the acknowledgment of receipt for some time, as simple network congestion could also be the cause of the missing ACK. However, in an attack, semiopen connections created by the malicious client consume resources on the server, and can ultimately exceed the available resources and cause a server service failure.

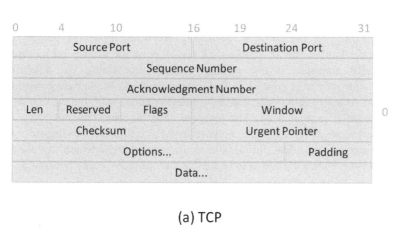

(a) TCP

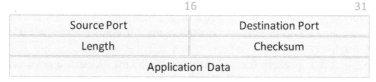

(b) UDP

Figure 2.15. *TCP and UDP frame*

2.4.3. Unified Datagram Protocol (UDP)

UDP is, like TCP, a transport layer protocol, but unlike TCP, it does not manage the notion of session and is therefore classified as stateless. In addition, it does not manage the loss of packet order and only implements the addressing of programs by ports. A typical UDP header is shown in Figure 2.15. UDP is mainly used for streaming services such as radio or Internet TV, but it is also the most widely used transport protocol for DNS. The advantage of UDP is its much higher speed than TCP.

2.4.4. Address Resolution Protocol (ARP)

Address Resolution Protocol (ARP) is used between layers 2 (Ethernet) and 3 (IP). It is used to establish the correspondence between MAC addresses and IP addresses. The opposite is done by Reverse Address Resolution Protocol (RARP). When a source host tries to communicate with a destination host for the first time, it sends a communication request to all workstations in the network, indicating the destination's IP address. The machine that has this IP address will be the only one to respond by sending an ARP response to the sending machine containing the information < IP_address, MAC_address >. To send this response to the right computer, it creates an entry in its ARP cache (translation table) from the data contained in the ARP request it has just received. The machine that made the ARP request receives the response, updates its ARP cache and can therefore send the message that it had put on hold to the computer concerned. An attack to intercept communications is to make this table corrupt (cache poisoning).

2.4.5. Internet Control Message Protocol (ICMP)

ICMP, documented in RFC 792, is a protocol that is closely integrated with operation of the IP. ICMP messages, delivered in IP packets, are used for messages related to network malfunctions. Of course, since ICMP uses IP, the delivery of ICMP packets is not reliable, so hosts cannot rely on receiving ICMP packets for any network problems. A well-known use of ICMP is made when sending an ICMP echo request with the ping command to test if a computer is reachable and to measure network latency. Other ICMP messages contain error information (non-accessible hosts, non-accessible network portion), congestion information (router receiving too many packets, timeout, etc.), or redirection information to indicate to a host that there is a better router to reach its destination.

2.4.6. *The IPv6 protocol*

IPv6 is the latest version of the IP. It is an improvement of IPv4 that has been developed to address the main deficiencies of IP4, including address depletion and security. IPv4 addresses have 32 bits (4 bytes): they change to 128 bits with IPv6, which is adapted to the development of the IoT. We have about 3.4×10^{38} addresses. To better understand the size of this address space, it can be seen that each human being will be able to have at least several billions of billions of addresses. An example of an IPv6 address is as follows:

4440:1023:AABA:0A01:0055:5054:9ABC:ABB0

Security is enhanced with the integration of IPsec (Annex 1) into IPv6, which means that two devices can automatically create a secure tunnel without any special intervention.

In addition, the header is improved, shorter and without checksum, making routing easier.

The ARP protocol, used by IPv4 to perform MAC address and IP address matching, is replaced by *Neighbor Discovery Protocol* (NDP) and *SEcure Neighbor Discovery* (SEND). One of the vulnerabilities allowing Man In The Middle attacks (Chapter 4) disappears.

The broadcast function (broadcasting a message to all devices) is deleted.

Finally, as there are enough addresses available, an address is associated with each device, and there is no longer any need for an address translation mechanism (NAT) and a private network concept as in IPv4, with addresses defined in the range 192.168.0.0 to 192.168.255.255, for example. For security reasons, however, IPv6 has an equivalent of the private IPv4 address in the form of a unique local address that is not globally routable. This way, the device is not visible from the outside.

Both IPv4 and IPv6 versions can coexist on the same network.

2.5. Industrial protocols

2.5.1. *Introduction*

The equipment of an ICS exchanges information using many protocols. Some are designed for specific application areas such as control systems in the chemical or petroleum industry, construction, electrical distribution systems, or for vehicles.

In general, in an ICS, it is first necessary to transmit data between sensors and actuators and level 1 processing systems. The transmission distance can vary from a few meters to a few hundred meters; the transmission can even be carried out over a much greater distance for telemetry systems. At this level, we talk about fieldbuses, and there are, for example, Modbus, Profibus, Asi-bus or DeviceNet protocols. The transmission is often made via an RS232 or RS485 link, a radio link or a GSM link. As a reminder, it can also be carried out by an electrical signal up to the PLC input and does not use a computer protocol (Chapter 1).

At the higher level, between levels 1 and 2, it is necessary to transmit information between the processing units (PLC, RTU, etc.) and the supervision system. The protocols used at this level are often Modbus, Profinet, Ethernet/IP or OPC-UA-based exchanges (where OPC is *OLE for Process Control and UA is* Unified Architecture). Data transmission is carried out via Ethernet, either in its classic non-deterministic version or in a so-called real-time version. Exchanges at higher levels (above level 2) are generally based on traditional computer protocols such as Ethernet or Wi-Fi and TCP/IP.

Since the 1990s, there has been a convergence of protocols that first used Ethernet 802.3, then wherever real-time aspects are not critical, use IP, which also allows the use of Wi-Fi equipment. IP network equipment is ubiquitous, easy to deploy, relatively inexpensive and well known to administrators. This technology facilitates the connection of manufacturing systems with the company's IT systems for better production management. It should also be noted that this convergence is not limited to ICS and business management aspects, but that voice (VoIP), television, video surveillance, and many services also use IP networks. It explains the significant increase in companies' vulnerability to cyber-attacks.

Fieldbus protocols that do not use TCP/IP or even Ethernet, as actuator/sensor interface (AS-i) presented below, are not vulnerable to attacks from the Internet[2]. Similarly, networks using an Ethernet protocol but not TCP/IP (such as Profinet RT) are, to some extent, less vulnerable.

---

2 Unless you can imagine a specific hardware device.

2.5.2. Modbus

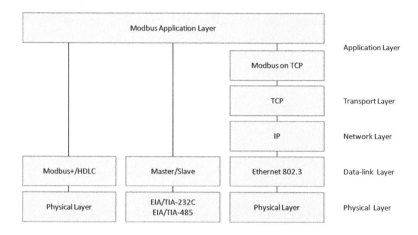

Figure 2.16. *Modbus and OSI communication stack*

Modbus is one of the oldest industrial control protocols. It was introduced in 1979 for PC-automated communications and used a serial type communication. In the 1990s, it grew considerably and, with a view to better integration with modern systems, a version for TCP/IP networks, called Modbus/TCP, appeared in 1999. At present, it is one of the most widely used protocols in many types of industries, including critical infrastructure. Modbus is a protocol located in the application layer (Figure 2.16), thus allowing different physical means of transport to be used. There are three types of Modbus implementation:

– Serial Modbus using RS232 or RS485 links, in RTU and ASCII version (bytes are encoded on two hexadecimal characters), using a variety of physical means of transmission (electrical cables, fiber, radio, etc.);

– Modbus+ over HDLC (high-level data link control) using a high-speed token-passing network;

– Modbus TCP/IP over Ethernet.

In general, the TCP/IP version is used between levels 1 and 2, that is, between PLCs and supervision, whereas the RTU version is used between levels 0 and 1 to connect PLCs and Intelligent Electronic Devices (IEDs) to PLCs. The trend is toward a generalization of TCP/IP even at the lowest levels.

The Modbus protocol is common to all implementations. It is a master–slave protocol. The master controls the communication, and the slaves respond to the requests. The protocol is very simple: there is no procedure to connect to the network, no detection of maximum response time and no diagnosis of the slave's status. The basic functions essentially allow words to be read or written from the slave's memory, and the frames contain a function code and associated data.

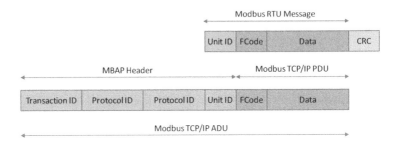

Figure 2.17. *Modbus frame. For a color version of this figure, see www.iste.co.uk/flaus/cybersecurity.zip*

The Modbus protocol is not secure, the frame content (Figure 2.17) is easily accessible, and they can be easily modified. In addition, since some commands can be issued in broadcast mode (sent to all receivers), a denial of service (DoS) attack can easily be performed. The vulnerability of Modbus is all the more worrying as it is used over TCP/IP.

2.5.3. *Profibus and Profinet*

Profibus (PROcess Field BUS) is a fieldbus communication standard promoted in 1989 by the German Ministry of Education and Research and an association of several institutions and industries including Siemens. It is based on serial communications by cable (RS-485) or optical fiber. It is a master–slave protocol like Modbus, but a little more sophisticated, with a procedure to join a network and time validations.

Profibus is a fieldbus mainly designed to connect field devices to the PLC and is not vulnerable to attacks via TCP/IP.

Profinet is a Profibus-based standard that adopts Ethernet as the physical interface for connections rather than RS485 and has a token-based repetition system to provide real-time functionality, modifying the protocol over a fraction of the time with a given periodicity, with the remainder of the period operating in conventional 802.3 Ethernet (Figure 2.18b). Profinet offers full TCP/IP functionality for data transmission, which allows applications to use it with a wireless network and high-speed transfers. The equipment using Profinet is oriented toward reliability and real-time communications. Figure 2.18a shows the architecture of Profinet.

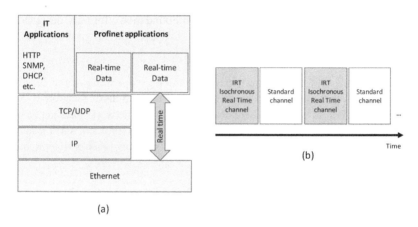

Figure 2.18. *(a) Profinet architecture; (b) Ethernet time division*

Profinet does not have any native security features. The functionalities incorporated in the protocol focus on improving system availability and operational reliability. To prevent potential attacks on a Profinet network, it is necessary to set up a segmentation of networks, Virtual LANs (VLAN) or introduce a DMZ (Chapter 10).

2.5.4. *Actuator/sensor interface*

AS-i was defined and developed by a consortium of 11 German and Swiss companies between 1990 and 1994. The objective was to provide a suitable fieldbus for connecting sensors and actuators. The system was designed on the basis of a simple unshielded two-conductor cable, capable of transferring data and a 24 V power supply. It uses a serial equipment topology and a master–slave protocol with a defined response time. On a bus, there is only one master and one power source. An AS-i message consists of a master request, a master pause, a slave response and a slave pause. By design, AS-i does not work over TCP/IP and is not one of the most vulnerable protocols.

2.5.5. *Highway Addressable Remote Transducer*

The Highway Addressable Remote Transducer (HART) protocol has been in existence since the late 1980s. It is a two-way communication protocol that provides data access between intelligent instruments and host systems. It is managed by HCF (HART Communication Foundation). Initially designed to operate with a 4-20 mA wired link, it has evolved to give rise to the WirelessHART protocol.

WirelessHART is a secure network technology operating in the 2.4 GHz radio band according to the IEEE 802.15.4 standard (section 2.3.4). Communication is managed between nodes using a network manager, and a security manager manages authentication. The HART protocol stack consists of four layers: physical layer, data link layer, transport layer and application layer.

Security services are offered at the data link level (level 2) and at the network level (level 3) in order to obtain a level of security equivalent to the wired solution. All messages are encrypted.

2.5.6. *DNP3 and IEC 60870*

DNP3 is a communication protocol developed in 1993, and widely used in the electricity sector, mainly in the United States and Canada. It is rarely used in Europe, as there are alternatives such as IEC-60870-5-101 or IEC-60870-5-104, which are different versions. It is a three-layer protocol, operating at the data link, application and transport layer levels. It was adapted in 1998 for use on an IP network with packet encapsulation with TCP and UDP.

DNP3 is a protocol designed to maximize system availability, with less emphasis on confidentiality and data integrity factors. At the data link level, some functions are included for detecting transmission errors by calculating the cyclic redundancy check . No additional security measures to the Ethernet protocol are proposed at this level. At the application level, a variant of DNP3 adds the possibility of authentication, which solves several security problems. It prevents identity theft and message modification by malicious actors.

2.5.7. The CAN bus

Controller Area Network (CAN) is a protocol that was developed by Bosch and is currently described by the ISO 11898 standard. It partially defines services for layers 1 (physical) and 2 (data link) of the OSI model.

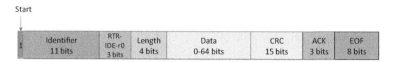

Figure 2.19. *CAN frame. For a color version of this figure, see www.iste.co.uk/flaus/cybersecurity.zip*

CAN transmissions operate on the producer/consumer model. When data are transmitted by a CAN device, it concerns another device designated by an identifier field. This ID field, which must be unique on the network, also provides the priority of the message. All other CAN devices listen to the sender and only accept messages about them.

This protocol is a serial type protocol. A CAN protocol frame consists of an identification field and a data field of up to 64 bits, plus various synchronization and verification bits. It is widely used in the automotive world: most of the models currently produced use it. It is relatively unsecured and has been the subject of famous attacks, such as the Jeep hacking in 2015 (Miller and Valasek 2015).

2.5.8. Ethernet/IP and Common Industrial Protocol (CIP)

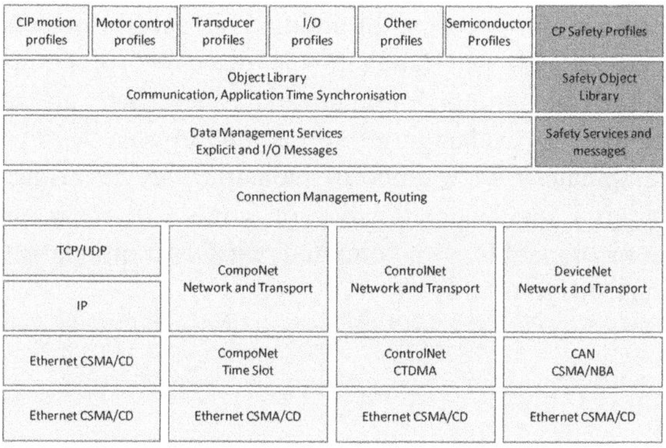

Figure 2.20. *Common Industrial Protocol*

The Common Industrial Protocol (CIP) is a protocol developed by the Open Device Vendor Association (ODVA) to automate industrial processes. CIP includes a set of services and messages for control, security, synchronization, configuration, information, etc., which can be integrated into Ethernet networks and the Internet. A number of CIP adaptations, providing intercommunication and integration for different types of networks, have been proposed:

– Ethernet/IP: an adaptation of CIP to TCP/IP;

– ControlNet: a CIP integration on Concurrent Time Domain Multiple Access (CTDMA)[3], which allows deterministic time communications for critical real-time applications;

– CompoNet: a version adapted to time-division multiple access (TDMA) technologies;

– DeviceNet: a CIP adaptation with a CAN controller presented above.

CIP is a protocol based on object-oriented modeling. Each object consists of attributes (data), services (commands), connections and behaviors (relationships between data and services). CIP offers a wide range of objects to cover typical communications and functions with common elements in automation processes, such as input or output devices, whether analog or digital, HMIs, motion controls, etc. The notion of profile makes it possible to define if a common communication object is implemented in a piece of equipment. The communication is based on the producer–consumer model, which can be single or multicast. In the first case, we are talking about explicit messages and in the second case, implicit messages.

Although CIP uses a formalized model, it does not define any mechanism, implicit or explicit, for security. In addition, it defines objects required to identify devices, which can facilitate the discovery of equipment in the network and provide targets for attackers. Since there are also common application objects for exchanging information between devices, an intruder is able to manipulate a wide range of industrial devices using this type of object. In addition, the characteristics of some CIP messages (real-time, multicast) are incompatible with communication encryption, so CIP does not provide mechanisms to do this.

---

3 To manage simultaneous transmissions, Ethernet uses the non-deterministic *Carrier-Sense Multiple Access with Collision Detection* (CSMA/CD) protocol.

Ethernet/IP (IP stands for Industrial Protocol) is a standard developed in the 1990s. EtherNet/IP is a CIP encapsulation for Ethernet networks. Commands, data and messages are provided in CIP frames.

As a CIP protocol, Ethernet/IP defines two connection methods for its TCP/IP communications. These are explicit messages, using the TCP, and implicit messages using the UDP. Explicit messages follow the client–server or request–response connection model. Among them are messages between PLCs and HMIs, diagnostic messages and file transfers. The commonly used ports are TCP port 44818 and UDP port 2222.

Although Ethernet/IP is more modern than protocols like Modbus, it still has security issues that leave it vulnerable.

2.5.9. *OLE for Process Control (OPC)*

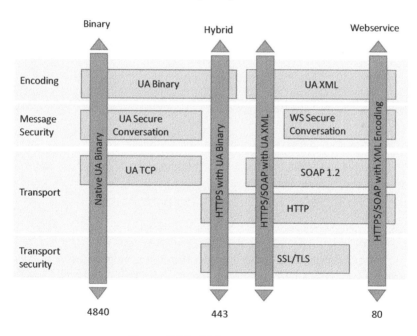

Figure 2.21. *OPC architecture*

OPC is not an industrial communication protocol, but rather an operational solution using object linking and embedding (OLE) for communications in Windows-based industrial control systems. OPC provides communication capabilities for ICSs using Windows workstations.

OPC was originally based on DCOM (Distributed Component Object Model) and many OPC systems still use it, although there is an updated version called OPC-UA (Mahnke *et al.* 2009), which allows the use of SOAP over HTTPS and is much more secure. This version has many open-source[4] implementations and can be used on different operating systems.

The use of DCOM and RPC (Remote Process Call) makes OPC very vulnerable to attacks and, as OPC is based on OLE, it can also be affected by all its vulnerabilities. In addition, since OPC is running on Windows, it can also be affected by all operating system vulnerabilities. As it is quite difficult to apply patches to industrial control systems, many vulnerabilities already discovered and for which patches are available continue to be exploitable on some industrial control networks.

OPC-UA has a security model (Armstrong and Hunkar 2010) that provides greater security to the architecture, so it is advisable to deploy OPC-UA rather than the traditional version of OPC.

2.5.10. *Other protocols*

There are many other protocols such as Powerlink Ethernet (another real-time version of Ethernet), EtherCAT (Ethernet for Control Automation Technology) and KNX (specialized for building). Details on these protocols can be found in Zurawski (2014).

2.6. IoT protocols

A number of protocols have been developed for the IoT. A first protocol, called 6LowPAN, is located between the network layer and the link layer of the OSI model, while others are at application level, such as Message Queuing Telemetry Transport (MQTT) and Constrained Application Protocol (CoAP), and are located at level 7 of the OSI model (Figure 2.11).

---

4 Available at: https://github.com/open62541/open62541/wiki/List-of-Open-Source-OPC-UA-Implementations.

2.6.1. *6LowPAN*

6LoWPAN, created by a group from the Internet Engineering Task Force (IETF), is an acronym formed from IP6 and low-power wireless personal area networks (LoWPAN). Its objective is to allow devices with limited processing capacity to transmit information wirelessly using Internet protocol. It is a competitor of Zigbee. It allows 802.15.4 devices to communicate using IP6. The main features of 6LoWPAN are as follows:

– packet fragmentation and reassembly, as 802.15.4 frames have only 81 bytes;

– IP6 header compression;

– the routing of fragments in the local network;

– functions for data security.

Examples of applications are smart meters. In addition, 6LoWPAN is very promising in home automation applications such as lighting or thermostats.

2.6.2. *Message Queuing Telemetry Transport*

MQTT is a publication/subscription messaging protocol for Internet devices on low-resource objects and low-bandwidth, high-latency or unreliable networks. Given its characteristics, it is an ideal protocol for machine-to-machine (M2M) communication.

MQTT was developed in 1999, but with the exponential growth of the IoT, and the need to connect and communicate low-powered intelligent devices, MQTT has recently found a market. MQTT was designed to be a low overload protocol, which took into account bandwidth and processor limitations.

MQTT is a publish/subscribe protocol. It allows customers to connect as a publisher, subscriber or both. Communication is ensured via a broker who manages all transmitted messages. This model allows customers to communicate in one-to-one, one-to-many or multiple-to-one mode.

Each message has an address, called a "topic". For example, the name of a topic can be `temperature`. Namespaces are structured in a tree structure such as:

```
plant1/compressor/valve1/temperature
```

Wildcards are allowed when registering a subscription (but not when publishing), which allows entire hierarchies to be observed by customers. For example:

```
plant1/compressor/+
```

allows you to subscribe to all compressor messages.

The operation is as follows: when the temperature value is changed in the "valve1" object, it issues a message to the broker. All subscribers receive a message with the new value.

2.6.3. *CoAP*

CoAP is the IETF CoRE group protocol (Constrained Resource Environments) for low-resource systems and can be seen as a light version of the HTTP protocol.

Like the latter, CoAP is a document transfer protocol. CoAP packets are much smaller than HTTP streams. Encoding is done to save space. Packages are simple to generate and can be analyzed without consuming a lot of memory. CoAP works on UDP, not TCP. Clients and servers communicate via stateless datagrams, and CoAP allows UDP broadcasting and multicast broadcasting. CoAP follows a client/server model. Clients make requests to servers, and servers send responses. Customers can read, modify, create and delete resources.

CoAP is designed to interoperate with HTTP and the RESTful Web in general because of simple proxies. As CoAP is based on UDP datagrams, it can be used over SMS and other packet communication protocols.

2.6.4. *Other protocols*

Advanced Message Queuing Protocol can be seen as an advanced version of MQTT, but it is less widely available. It has been developed by the banking world to ensure secure transactions between servers; it is adapted for IoT, in particular to meet the requirements of critical applications.

DDS is a Data Distribution Service for real-time systems defined as standard by the *Object Management Group*. It is suitable for M2M exchanges and allows real-time data exchanges that are reliable, efficient and interoperable. It works in publish/subscribe mode. DDS meets the needs of applications such as financial trading, air traffic control, smart grid management and other Big Data applications.

XMPP, Extensible Messaging and Presence Protocol, are a set of open client/server protocols for IETF instant messaging. It is used by Apple, Cisco, Microsoft and many industry players. It can be used as part of an IoT architecture between a platform and HMI.

3
IT Security

3.1. Security objectives

The first part of this chapter introduces what are known as Information Technology (IT) security objectives and details them in the context of the industrial control system (ICS). The second part gives useful, precise definitions concerning the notion of risk. The principle of risk analysis, based on the assessment of impacts and their likelihood, is then described. The last part presents the evaluation process and shows how it fits into the PDCA continuous improvement process (Plan, Do, Check, Act).

The different methods of risk analysis are presented in Chapter 9. The general risk management approach for ICS is detailed in Chapter 11.

3.1.1. *The AIC criteria*

In the world of information security, we often consider three properties of a system that must be guaranteed: the availability (A) of services, the integrity (I) of software and data and the confidentiality (C) of information. These criteria are called AIC criteria. They are seen as objectives to be achieved by security functions.

The importance of the first criterion, the *availability of* a service or information, is quite easy to perceive. Everyone can see this when they need a web service from their workstation or mobile phone.

This criterion is more or less important depending on the context and the user: the lack of an available mapping service used to locate a destination mentioned in a documentary will be less problematic than it will be if one is lost in a city, abroad and at night.

With regard to the availability of information services for industrial or cyber-physical systems, a distinction must be made among:

– real-time or near real-time control functions;

– safety functions;

– functions for analyzing and consolidating production data in relation to the IT world.

A lack of availability of the functions from the first category can lead to a shutdown of the system and therefore of production. This is quite similar to traditional computer systems. For the functions in the second category, a lack of availability can lead to a loss of control and put at risk the safety of people and assets. The damage can be very significant. For the last category, functions for analyzing and consolidating production data, we find ourselves in a situation similar to that of a traditional information system (IS).

Adequate availability is ensured by a design that combines reliability and redundancy of equipment, on the one hand, and adequate security measures, on the other hand. In the event of unavailability, a business continuity plan can guarantee service continuity, possibly with degraded quality.

The second criterion concerns the *integrity of* data and software. The objective is to keep them intact, whether against malicious acts, human errors, material failures or natural phenomena. A loss of integrity can result in the destruction of information or partial modification.

Data integrity is a fundamental aspect. Regardless of malicious intent, this must be ensured to avoid any corruption of information.

The traditional mechanisms used are integrity checks, duplications or backups and redundancies of storage systems (such as a RAID disk) or processing systems (redundant units with voting). Data integrity can be proven by digital signature mechanisms (Appendix 1).

In the case of ICS, integrity also concerns industrial equipment, that is, its configuration or programs executed by it and data present in the memory.

The integrity of different types of data can be considered:

– configurations and programs of devices such as Programmable Logic Controllers (PLCs) or acquisition systems;

– Supervisory Control and Data Acquisition (SCADA) and historian system configurations;

– the data stored in the PLC memory, updated in real time and representing the status of the cyber-physical system;

– permanently stored data, which comes from measurements of the parameters of the cyber-physical system (historical).

The third criterion is *confidentiality*. It is linked to the maintenance of secrecy and is obtained by a reading protection. Confidentiality is related to authentication and access rights management to control who can read data. It can also be obtained by encrypting data using a cryptographic algorithm (Appendix 1).

A complementary criterion is often used; this is the evidence of evolution (E) that includes several aspects:

– *traceability*: this is ownership of an asset making it possible to find with a given level of confidence the circumstances in which the asset evolves (origin and successive actions);

– *accountability*: this is the property that characterizes, with a given level of confidence, assignment of an action to a given entity (machine or person);

– *non-repudiation*: this is a property associated with an action or event that ensures it has taken place and it cannot be denied later. For example, the impossibility for a person or any other entity involved in a computer communication to deny having received or sent a message.

This criterion is not always taken into account; it is not as important as the first three and it is quite difficult to put in place mechanisms to track and prove the addition, deletion or modification of information.

For ICS, traceability is fundamental for certain applications in the world of pharmacy or health, for example.

To ensure the security of a system and its data, additional functions such as *identification* and *authentication are* also used to ensure that only rights holders can access a service or data, and that they can modify them only if they are authorized to do so. These additional identification and authentication functions ensure:

– the integrity of the data, with regard to unlawful modifications, and their confidentiality, since only authorized persons may have access to these data and modify them if they have the right to do so;

– accountability, since the actions carried out are carried out by identified persons.

Mastering the security of a computer system therefore consists of ensuring that these properties are satisfied, with a certain level of confidence.

The AIC security needs for an ICS are not completely identical to those of a management system, particularly in terms of priorities. Indeed, for an IT system, confidentiality is often a priority, and this is not the case for an ICS (Figure 3.1). In addition, the criteria of integrity and availability are expressed in different ways. The integrity of programs and data ensures the safety of people and property, on the one hand, and the performance of the system, on the other hand. Availability, on the other hand, ensures security for the Safety Instrumented System (SIS) and performance and quality for the Basic Process Control System (BPCS).

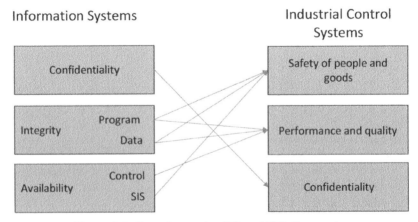

Figure 3.1. *Security needs of IS and ICS. For a color version of this figure, see www.iste.co.uk/flaus/cybersecurity.zip*

3.1.2. *The different levels of IT security*

The purpose of information system security (ISS) is to protect data. It can be broken down into a series of levels that must be reached by anyone wishing to damage the data.

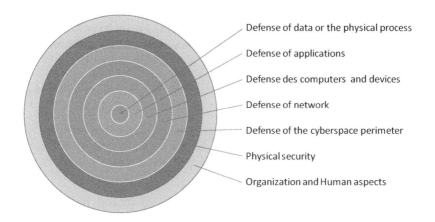

Figure 3.2. *The different layers of IT security. For a color version of this figure, see www.iste.co.uk/flaus/cybersecurity.zip*

The most external level concerns the organizational and human aspects, on which the rest is based. At this level, responsibilities must be defined, an appropriate policy put in place and human stakeholders sensitized and trained.

The second level from the outside concerns the physical security of the system. To ensure this, it is necessary to provide the following:

– protection against physical access to premises or equipment to prevent damage, theft or the installation of material intended for piracy;

– protection of equipment against natural phenomena such as fires, floods, earthquakes, overheating, etc.;

– protection of equipment against failures of the energy supply system and, where applicable, cooling;

– the reliability of the various equipment or networks used, which is ensured both by the quality of the equipment and by an appropriate preventive maintenance policy;

– possible redundancy of critical equipment.

Perimeter defense in cyberspace is about protecting the IS from threats from outside. These threats are targeted actions or systematic actions attempted on all elements connected to the Internet. As these attacks can be carried out by a wide variety of particularly active actors, they represent a considerable and constantly increasing risk.

Network defense concerns threats related to the transport of information. Vulnerabilities are very diverse. There are different types of attacks:

– network listening: this consists of copying the traffic of a network. This type of attack does not generate direct damage to the system, is difficult to detect and results in the theft of information that can be reused to conduct other types of attacks;

– session theft: the attacker waits for a user to connect remotely with his/her login and password, and then takes control of the connection;

– identity theft: this consists of using another user's MAC or IP address;

– connection hijacking: this consists, for an attacker, of interposing himself between the source and the destination of a connection. This attack is also called a "Man-in-The-Middle attack";

– denial of service: this consists of flooding a network machine with a large number of messages in order to render it inoperative.

It should be noted that some attacks are facilitated by physical access to the network and easily accessible RJ45 sockets are important vulnerabilities.

Defense of computers focuses on direct attacks on machines and operating systems (OSs). Attacks via USB ports or devices connected between the keyboard and the CPU are to be taken into account.

The machines are managed by an OS that offers protection mechanisms such as authentication mechanisms, resource access control mechanisms, cryptographic mechanisms and even a firewall. It is also possible to install antivirus and antimalware software.

At the same time, the OS is also the target of attacks and presents vulnerabilities. The most serious are those resulting from the design or implementation of the OS, which have been discovered, but which have not been the subject of any patch publication. The existence of such a vulnerability can be seen as an open door to the system and protection should be based on the other levels.

Note that in the world of ICS, devices can work with specific OSs such as VxWorks, QNX or even proprietary systems for which vulnerabilities must be remedied.

Application security is about ensuring that applications do not offer vulnerabilities, similarly to OSs. These applications must provide mechanisms to authenticate users and prevent unwanted actions on different devices and data.

Data security concerns the security of information, both in memory and on storage devices. The data can be software or configurations of different devices and are therefore critical. OSs offer authentication and rights management mechanisms that, coupled with a security management policy, enable data security to be controlled. Cryptographic algorithms are used to enhance the level of security.

In the world of ICS, protocols and devices often provide relatively easy access to data and special measures must be taken to protect the data, as explained later in this book.

It therefore appears that, in order to corrupt a computer system, it is necessary to go through a number of successive layers. We find the classic model encountered in risk analysis (Figure 3.3) (Flaus 2013). This model is at the origin of the defense-in-depth strategy that consists of adding layers in the form of protection and reducing the faults of each level in order to limit as much as possible the possibility for an attack to cross all levels.

This defense-in-depth strategy is particularly relevant in IT, where most protection systems have unknown intrinsic vulnerabilities, which can be discovered by attackers and traded in hidden markets (zero day vulnerabilities).

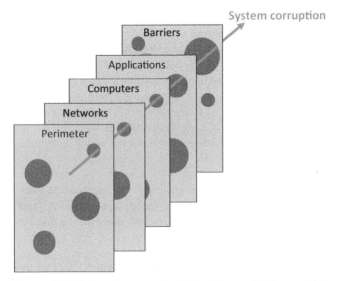

Figure 3.3. *Swiss cheese adapted to IT security. For a color version of this figure, see www.iste.co.uk/flaus/cybersecurity.zip*

3.2. Differences between IT and OT systems

To analyze the differences in cybersecurity between the OT and IT worlds, we will consider four aspects:

– the expected functionalities;

– the technology used;

– the lifecycle of the system;

– security management.

3.2.1. The *functionalities*

An ICS is a system used to monitor and act on a physical system. This is to be compared to an IT system, which only handles data and only operates in a digital world.

The main functions of an ICS are, on the one hand, to ensure optimal control according to the planned operating mode and, on the other hand, to provide safety functions in the event of abnormal operation.

As part of normal control (BPCS), the industrial control-command system provides the functions for acquiring and storing data from the sensors, regulates the parameters to the desired values and controls the physical system according to predefined sequences to carry out production.

As part of the safety functions (SIS), the ICS's mission is to detect deviations that may affect the security of the physical system or its environment, and to carry out actions to secure the system. It can, for example, detect temperature or pressure thresholds, or dangerous positions, and open a valve, trigger cooling or block an action.

The response time of an ICS is closely related to the real time of the physical world: we are talking about real time or near real time systems. It is clear that availability, i.e. the ability to perform a function at a given time, is an essential feature of ICS, particularly for security functions.

For example, a Web server can, from time to time, operate with response times of 1 s. For a system operating a vehicle, this is much more problematic: at 60 km/h, 16 m are covered per second.

3.2.2. *The technology*

Unlike IT systems, which are mainly composed of computer workstations, servers and network equipment of controlled and relatively recent quality, ICSs are based on a variety of equipment that can be built from standard modules (COTS) or be quite old. The various field equipment directly connected to the physical system are mainly PLCs and embedded systems. These are designed primarily to provide efficient real-time functions and offer limited support for security or centralized administration.

These limitations may result from an old design, a too low embedded computing capacity or a limited energy autonomy. For these reasons, it is not simple to implement secure solutions such as encryption or authentication based on a global directory.

At the communication level, an ICS uses different network types and protocols (Chapter 2). Some protocols are particularly insecure.

This technological heterogeneity creates significant vulnerabilities. In addition, the deployment of security updates can be complex.

3.2.3. *System lifecycle*

An ICS is designed during the design of the installation. Its lifespan is identical to that of the installation and can last up to 20 years (Stouffer *et al.* 2015).

As a result, the system cannot evolve in response to changing threats and incorporate the latest security measures. This is the case, for example, with data encryption between devices, which cannot be easily added.

In many cases, ICSs operate with remote equipment: this is the case for electricity networks, petroleum product transport networks, water treatment systems or dam monitoring. This geographical dispersion poses a security challenge: remotely operated systems can experience cyber-attacks via the communication channel and are vulnerable to physical attacks that allow data, passwords or cryptographic keys to be stolen.

As we have said, an ICS is a complex system and, more often than not, equipment comes from different vendors, with vulnerability management policies that may be different and not very responsive.

In addition, an ICS must operate 24 h a day, 7 days a week and few facilities have a test platform. Security updates are therefore difficult to deploy. This compares to the approach of large software vendors, for whom security is crucial, and who provide security patches quickly.

To install an ICS, the operator often relies on an integrator to deploy, configure and program the system. This means that he does not always have a detailed technical knowledge of his system, either in terms of equipment or communication.

In operational mode, ICS management requires several types of skill, as IT equipment is managed by personnel other than the employees managing the Operation Technology (OT) system.

There are cases where the IT part of the ICS is managed by the IT department independently of the OT part, itself managed by the production or maintenance staff, and other cases where the ICS is managed globally and independently of the IT system by the production staff.

In all cases, there is a division of responsibilities between several departments that is not always clear, and a need for multiple competencies between internal and often external stakeholders.

3.2.4. Security management

Industrial installations generate risks for employees, the environment and populations. In addition, there are risks of the production equipment malfunctioning or breakdown, which can have an impact on the quantity or quality of production. These risks are generally the subject of a detailed risk analysis, and are even mandatory for high-risk installations (such those covered by the Seveso directive), in order to ensure what is called "the operational safety of the installation" (Chapter 8).

To manage these risks, a security management system is used in many companies. In most cases, it does not support IT security, which remains the responsibility of the IT department. In principle, the information security control approach is based on the implementation of good practices and does not systematically use risk analyses.

Risk management between the IT and OT world is therefore not generally carried out in an integrated way.

A first consequence is that, very often, even if the actors have a good industrial risk culture and are informed of the risks related to IT security for the IT part, their perception of this risk for the OT part remains rather limited, even if it is improving.

This is reflected, for example, in the management of access rights. In a traditional computer system, we usually distinguish between users with more or fewer privileges and administrators. In an ICS, the distinction is rarely based on the person, but rather on the role: SCADA operators working in 3 × 8 shifts often share the same authentication. In addition to administrating the computer system itself, it is also necessary to make adjustments to PLC parameters or program modifications. In many cases, the development task is outsourced to an external company, which may have to intervene on the system put into production, and often remotely. Very often, these accesses are not well controlled in terms of security.

Both the definition of rights and the authentication, which is classic in the field of IT, are therefore less systematic.

Finally, it should be noted that implementation of basic security measures such as antivirus software is almost systematic on IT systems, and that their updating is automatic. This is not always the case for workstations used in ICS. Indeed, on the one hand, the functioning of the antivirus and its update may be incompatible with real-time needs and, on the other hand, access to an update server is not always possible from the OT network.

3.2.5. IT/OT convergence

The previous sections have shown the differences between the IT and OT parts. However, the current trend is toward the convergence of these two areas (Pettey 2017). The objective is to reduce costs and improve functionality.

This convergence is made possible by technological developments, which make it possible to distribute processing capacity and improve communications. Industrial installations are moving toward the Internet of Things (IoT) for interaction with the physical system, with control of the production line via IoT equipment. This is also the case for energy management, or in the concept of smart cities with distributed sensors. In these architectures, data are stored in the cloud, which also offers data processing and exploitation capabilities (Chapter 2).

Interaction with the user also takes new forms in this context: it involves mobile equipment, often using users' personal terminals, and is in line with the BYOD (Bring Your Own Device) trend.

These new architectures (industrial IoT, cloud computing and mobile computing) have significant impact on security: some aspects are positive, such as the use of better secured protocols or better rights management, others are negative, such as the use of terminals for which the guarantee of integrity is low, or the geographical distribution of equipment with its data, which increases vulnerability.

3.2.6. Summary

Table 3.1 summarizes the IT/OT differences. As we can see, these differences are significant, which has important security consequences:

– the impacts of the risks are different, as well as what causes them;

– the operating constraints are different;

– the staff in charge come from different cultures and have different skills.

Security management must take into account all this heterogeneity.

	IT	OT/ICS
Security properties		
Generally accepted order	Confidentiality, integrity, availability, evidence	Availability, integrity, evidence, confidentiality
Availability	Important, but not critical	Critical for safety and production continuity
Integrity	Critical	Critical
Evidence of evolution	Important for some applications	Important for applications requiring traceability
Confidentiality	Critical	Not very critical
Technology		
Respect of response time	Desired (QoS)	Imperative, especially for security functions
Type of technology	Standardized and limited in number	Heterogeneous systems, many protocols, embedded systems and COTS
Exploitation		
Lifespan	3–5 years	May exceed 20 years
Update	Automatic	Tricky updates: – continuous operation – not always automated
Outsourcing	Usual in a global way	For design and integration and often maintenance
Security management		
Risk analyses	Global (based on security needs)	Detailed (based on the functions of the physical system)
Authentication and access rights	At user level	At role level, no global directory
Security awareness	Good	Awareness of risk is still relatively low
Antivirus	Almost systematic	Not always implemented or up to date

Table 3.1. *IT/OT differences*

3.3. Risk components

The definition of risk involves a number of related concepts presented in Figure 3.4 (Louis *et al.* 2016).

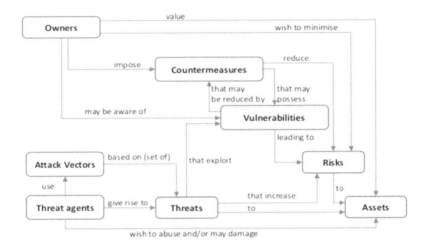

Figure 3.4. *Links between concepts related to risk*

3.3.1. *Asset and impact*

In IT security, the term asset is used to refer to the entities considered in the analysis, particularly those that are the target of a risk.

Impact is the effect or consequence of an undesirable event, characterizing a risk, on a human or informational physical asset, or on the environment. An impact can be more or less significant and is characterized by a level of severity.

For example, the impact of a server failure, following a computer attack, could be an unavailability or loss of data.

In general, for an IS, we can identify the following:

– impacts on production: loss of service production capacity or of products as a result of a technical malfunction of an IT system component, a communication failure, a denial of service or a loss of data;

– direct financial impacts: changes in invoice amounts, processing errors and unwanted transfer orders;

– impacts on competitiveness caused by theft of customer data or know-how;

– impacts on image and reputation;

– legal and regulatory impacts: questioning in the event of fraudulent use of customer data or company resources, non-compliance with obligations (operators of vital importance, LPM).

For ICS, the impacts will be mainly related to the production apparatus (in the broad sense) or the physical system, and we will have the following:

– impacts on production, whether in quality or quantity;

– impacts on property, people and the environment related to a loss of control of the physical system, whether it is a Seveso facility, a medical device, a train, a car, etc.;

– impacts on the service provided for systems for water distribution, energy, telecommunications, etc.

3.3.2. *Threats*

A threat is a phenomenon, thing or person that can potentially cause damage. A threat is generated by a source of threat or an actor called an "attacker".

An IT threat concerns the IS, that is, central units, the communication network or information storage systems. As part of the security of ICS, it can also affect various industrial equipment, such as PLCs.

The term "cyber" refers to cyberspace, and therefore in principle refers to threats from the Internet. By abuse of language, we often speak of cyber-threats for all threats to the ICS that are related to IT aspects.

ICSs are also vulnerable to hardware failure, natural phenomena, human error or direct human malevolence.

3.3.3. *Attacks*

Attacks are particular threats: they are malicious actions designed to compromise the security of the IS.

Attacks may target the availability, confidentiality or integrity of information.

These attacks can be active or passive, i.e. send information or commands to the IS, or simply observe the flows.

Attacks can be conducted by attackers of various levels. These are often classified according to their capacity:

– low capacity: for example, an amateur hacker;

– substantial capacity: for example, a hacker who is a member of a group;

– almost unlimited capacity: for example, a government agency.

It may also be useful to differentiate between attacks from outside (cyberspace) and inside the company, whether they are carried out in a way that is conscious (corrupt employee, disgruntled, outsider) or not (e.g. phishing).

3.3.4. *Vulnerabilities*

A vulnerability or flaw is a weakness in an IS that allows an attacker to compromise the integrity of that system, i.e. its normal functioning, or the confidentiality or integrity of the data it contains.

It may be due to a cause related to the design, comissioning, configuration, operation or maintenance of the asset.

Vulnerabilities can be technical: this family includes all vulnerabilities related to the use of technologies or solutions, whether hardware or software. Searching for vulnerabilities is a very active area and, as a result, new vulnerabilities appear daily.

Vulnerabilities can also be human. They are based on human feelings, behaviors and instincts. These behaviors are very often exploited in different attacks belonging to what is called "social engineering". These vulnerabilities include fear (often exploited by threats such as scareware), pity, curiosity, libido, greed, etc.

3.3.5. *Definition of risk*

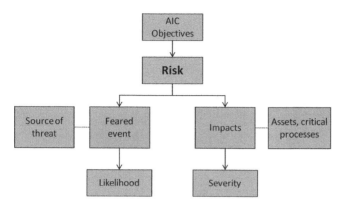

Figure 3.5. *Components of risk*

In the context of ISO31000, risk is defined as the effect of uncertainty on objectives. This definition can be reformulated for computer security or operational safety as an uncertain event that can have damaging consequences for a system. It is characterized by a likelihood or probability of the undesired event, and by a severity of the impact of the consequences.

In the context of information security, unwanted events are created by different types of threats, including attacks or cyber-attacks.

To analyze a risk, one must therefore consider the possibility of the feared event and the impacts it may have on the processes affected. The feared events are caused by the realization of threats, which is more or less likely, and have an impact on the goods and processes concerned, characterized by a level of severity (Figure 3.5).

To illustrate these different definitions, let us take the example of a ransomware distributed in an SME (small and medium-sized enterprise) via a malicious email and which affects the accounting workstation. The feared event is the activation of ransomware. The direct consequence is the encryption or deletion of data on the workstation, which can be expressed as a loss of availability of the accounting system. The threat is the trapped email, and the realization of it is its reception by the accounting post.

It should be noted that it is often easier to determine the severity of the impacts associated with a feared event than the event's causes and likelihood.

3.3.6. Scenarios and impact

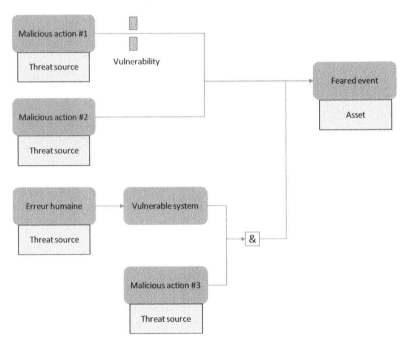

Figure 3.6. *Example of a scenario*

A scenario can be defined as a sequence of events that lead to a feared event that has an impact on the system. In this sequence, some events can also be combined with each other. The notion of the scenario makes it possible to describe in detail the conditions of occurrence for the feared event. A scenario may be more or less likely.

The impact of a risk can be defined as the effect of the feared event reflecting the risk on critical assets or processes.

For an ICS, the effects of a feared event can be classified according to whether they can lead to one of the following (Figure 3.7):

– a malfunction in the production process, and therefore a loss of service;

– a malfunction of the safety system (SIS), which may go unnoticed until the SIS is called upon, at which time it leads to a dangerous evolution of the system;

– dangerous evolutions or actions in the physical system caused directly by the sending of incorrect orders and which may lead, in addition to a loss of service, to damage to property and persons.

The significance of the impact may be different depending on the AIC component considered. For example, for a PLC, the loss of availability could result in a shutdown of the installation (impact on proper operation), whereas the loss of data integrity could result in damage to the controlled machine (property security).

All the impacts on the IS and the production system result in impacts at the global level of the company, most of which lead to financial losses.

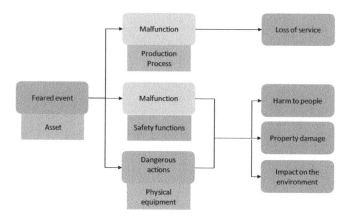

Figure 3.7. *Different types of impacts. For a color version of this figure, see www.iste.co.uk/flaus/cybersecurity.zip*

3.3.7. Risk measurement

To measure a risk, the importance of each of its components, i.e. the likelihood of the feared event and the severity of the impact, is assessed and then the level of risk is derived from this value pair. It is possible to select a qualitative or quantitative approach:

– with the first approach, a number of levels are defined by linguistic values, for example, low, medium, high, whether for likelihood or impact. Each value pair is then associated with a level of risk;

– in the case of the quantitative approach, we use a probability, number p between 0 and 1, and a numerical impact, noted i, which may be an assessment of financial losses in a given currency, or the number of people affected by the damage, or any other relevant indicator. The level of risk is then obtained by the product p × i.

In the case of an IT security risk analysis for ICS, the most realistic and effective approach is the qualitative approach. The quantitative approach, which seems less subjective, is difficult to implement because there is little data available and, due to this problem, it is not more objective after all.

To improve the repeatability of the evaluation, it is strongly recommended to detail the meaning of each level. The number of levels chosen usually varies from 3 to 5. Definitions such as those given in Tables 3.2 and 3.3, where each level is defined in detail, can therefore be found.

Level	Meaning of the word
Very unlikely	Occurs less than once every 100 years
Possible	Occurs less than once every 10 years
Occasional	Occurs less than once a year
Frequent	Occurs more than once a year

Table 3.2. *Likelihood levels*

Level	Availability	Integrity	Confidentiality
Minor	Less than a few minutes	A file altered but correctable	A file exposed internally
Serious	Less than a few hours	An altered file	Some files exposed internally
Major	Less than a week	Some altered files	Less than a few files exposed externally
Critical	More than a week	All files altered	All data exposed externally

Table 3.3. *Severity levels*

To allocate a level of risk using the couple (likelihood, severity), the most flexible solution is to use a risk matrix (Figure 3.8), also called heat map.

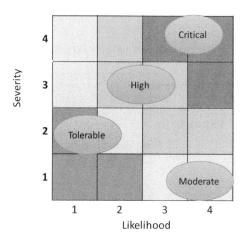

Figure 3.8. *Risk matrix or heat map. For a color version of this figure, see www.iste.co.uk/flaus/cybersecurity.zip*

Level	Meaning of the word
Tolerable	Accepted
Moderate	Accepted and monitored
High	Action required and planned
Critical	Immediate action required

Table 3.4. *Risk level*

3.4. Risk analysis and treatment process

3.4.1. *Principle*

The general scheme of a risk analysis for ISS is shown in Figure 3.9. It is detailed in Chapter 9. The main steps, after establishing the context and defining the scope of the analyzed system, are as follows:

– risk identification;

– the estimation of the level of risk through, in general, the characterization of the level of impacts and the likelihood of scenarios;

– risk assessment to determine whether or not the risk is acceptable.

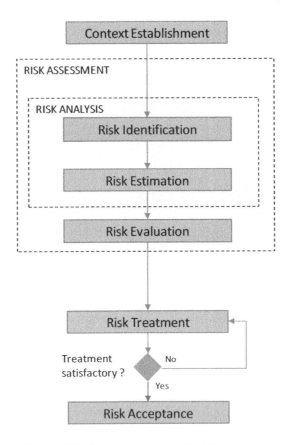

Figure 3.9. *General approach to risk analysis*

Depending on the level of risk, different types of treatment exist:

– the risk can be accepted;

– the risk can be transferred through an insurance mechanism;

– the risk may be refused, resulting in a shutdown of the installation or a modification of the infrastructure.

Most often, the aim is to reduce the risk by implementing measures that can affect the likelihood or severity component.

The choice between the different options depends on the level of risk and the organization's strategy.

In the latter case, technical and organizational measures are implemented. For example, in the case of a risk of attack on network integrity, encryption measures may be coupled to an intrusion detector.

As we will show below, it is often preferable to put different levels or layers of measures in place to protect against a risk. This is particularly true for information security, where some equipment or software may have design defects that leave doors open in the system.

3.4.2. Acceptance of risk

A risk is accepted when its level is low, i.e. when the likelihood-gravity pair leads to a criticality level decided as acceptable. This level can be defined via a risk matrix. In the example in Figure 3.8, this may be the tolerable or tolerable and moderate level.

When the risk is moderate, if the cost to reduce it is too high or there are no means to do so, one solution may be to accept it.

3.4.3. Risk reduction

Risk reduction is achieved by implementing a number of measures that address technology, human and organizational aspects.

These security measures can be classified according to several criteria. First of all, they can be distinguished according to the layer (Figure 3.10) they protect: physical security, perimeter security, etc.

It is also possible to distinguish between *technical* measures (e.g. firewalls), *organizational* measures (e.g. backup procedures) and those related to *human* skills and behavior (e.g. approach to deal with emails).

Finally, the measures can be classified according to where they are located in the chain of events leading to the damage:

— the measure may aim to prevent the occurrence of the undesired event, either:

- by intervening at the design level (secure architecture); or

- by implementing actions or mechanisms at the operational level (e.g. antivirus);

— the measure can seek to reduce the risk by detecting the occurrence of abnormal events as early as possible (as with an intrusion detector);

— the measure can aim to reduce the impact, by modifying the system (redundancy of a server on two physical sites, for example);

— the measure can facilitate recovery (backup, for example).

Lists of such measures classified by theme are available in ISO27002 (Chapter 6) for IT systems or in IEC62443 (Chapter 7) for ICS. Choice and organization of these measures must be carried out in such a way as to limit costs and maximize efficiency.

3.5. Principle of defense in depth

The principle of defense in depth comes from the military world. It appeared with Vauban, who, in the 15th Century, proposed the idea of low fortifications to resist metal balls capable of destroying vertical fortifications. In principle, the loss of a line of defense should weaken the attack or provide information (this condition is mainly specific to the military world).

In the world of risk management, we find the idea for hazard studies in industrial installations, particularly with the Layer of Protection Analysis (LOPA) method (Flaus 2013), and for nuclear safety.

The general idea of the two approaches is similar: multiple levels of protection, or lines of defense, introduced at the design stage of the installation, reduce the risk of an accident having serious consequences outside the installation to an extremely low level. Each security device, which is vulnerable and may be faulty, is added to those of the other layers to ensure security. The fundamental assumption is that each level must be independent, i.e. that the failure of one level must not be linked to the failure of the others.

In the world of information security, the idea is transposed by basing protection on a set of measures implemented at each level.

For each of the components listed in section 3.1.2, Figure 3.10 presents the measures that can be applied:

– organization: training, awareness, risk management;

– physical security: physical protection, locks, tracking devices;

– perimeter: firewall, quarantine, VPN;

– network: network segmentation, IPsec, Network Intrusion Detection System (NIDS);

– machines: OS security enhancement, regular updates, authentication, Host-based Intrusion Detection System (HIDS);

– applications: antivirus, configuration;

– data: access control list, encryption.

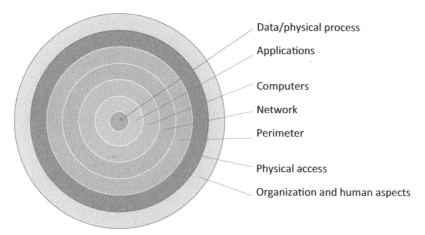

Figure 3.10. *Defense in depth. For a color version of this figure, see www.iste.co.uk/flaus/cybersecurity.zip*

In the context of ICS security, this approach is detailed in Chapter 11.

3.6. IT security management

Risk analysis makes it possible to put in place appropriate security measures. Risk management makes it possible to keep them as they are and to adapt them to the changes in the installation and to the threats.

Risk analysis has as an objective to assess the current situation with regard to security, and the choice of security measures makes it possible to define actions to be taken. A security management system integrates these steps into a process that manages the implementation of actions, evaluates their effectiveness and ensures that the approach is sustainable, and takes into account changes in the installation and the risk.

An Information Security Management System (ISMS) aims to control risks related to the IS, which results in the folllowing:

– optimal operation of the IS and ICS to guarantee a desired level of productivity or service quality;

– an acceptable level in terms of likelihood and severity for damage to the ICS and the production system;

– an acceptable level in terms of likelihood and severity for the damage generated by the production system to its wider environment (personnel, population, natural environment, rest of the company, etc.).

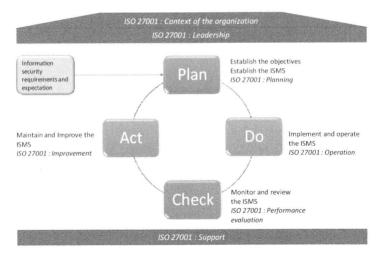

Figure 3.11. *Continuous improvement PDCA and ISO 27001:2013. For a color version of this figure, see www.iste.co.uk/flaus/cybersecurity.zip*

Such a process is formalized in various standards, including ISO 27001 (presented in Chapter 6). It is described in Figure 3.11. It is based on the PDCA standard, which is a method of continuous improvement. It should be noted that in the 2013 version, the explicit reference to this approach has been removed from the standard ISO 27001, but the idea of the approach is retained.

The step prior to implementation of the wheel is a step that consists of defining the context, similar to the first stage of the standard ISO 31000 (Flaus 2013). The specifics include the following:

– determination of the stakeholders affected by information security and their requirements. This part may include legal or regulatory aspects, such as the NIS Directive (Chapter 6), or contractual aspects;

– definition of the ISMS perimeter, considering the physical limits, organizational limits, data involved and, in the case of an ICS, the controlled system.

This preliminary step also defines leadership, which includes the following:

– governance and management commitment;

– security policy;

– the definition of roles and responsibilities.

The PDCA cycle can then begin. It is not specific to information security but is a general continuous improvement approach, proposed by W. Shewhart in the 2000s and then taken up by W. Edwards Deming. It may, for example, concern quality, environmental management and staff security. Its principle is as follows:

– before you start implementing something, you need to know exactly what you really need, and you need to know exactly what you want to achieve: this is the *plan* phase;

– once the objectives are known, the measures decided upon in the plan phase can be implemented: this is the *do* phase;

– however, the effort does not stop there. We must ensure that we have achieved what we planned, so we must monitor the system and measure whether we have achieved the objectives: this is the *check* phase;

– finally, if we realize that what we have achieved is not what we expected, we must close the gap. It is also possible to seek to improve the measures. This is the *act* phase.

In the context of industrial IS security, these phases are divided as indicated below.

In Phase 1, the Plan phase, which consists of carrying out the planning of the industrial security management system, there are four stages:

– In the first step, taking into account the context and scope, it is necessary to determine the risks to be considered in order to:

- ensure that the management system achieves the expected results;
- reduce negative impacts on the system;
- implement a continuous improvement approach.

– The second step describes the risk identification method (similar to the methods presented in Chapter 9, but the choice is free). Management must validate this method. This is then implemented to provide a list of risks with their assessment in terms of impact and likelihood.

– The third step consists of defining and implementing treatment of the risks identified in the previous step. This results in a list of security measures to be implemented, which makes it possible to establish a treatment plan. This plan must be approved by the parties concerned by the risks, and the residual risk must be accepted by them.

– The last step is to define the security objectives and the planning to achieve them. Security objectives must be measurable and take into account risk analysis and associated processing. They are the subject of a communication. Planning should specify the tasks to be carried out, the resources, the person responsible, the deadline and the measurement of results.

The advantage of this approach, driven by risk analysis, is that it allows efforts (human resources, budget, communication) to be focused on the essential aspects.

The risk analysis may have been carried out in the preliminary step to develop the security policy.

Phase 2 is Do. The action plan for implementing the security policy is deployed in this phase.

These actions are monitored: the implementation of the measures is documented and maintained. If certain actions are subcontracted, they must be subject to controls. At the same time, residual risks should be monitored as far as possible.

Phase 3 is the Check phase, which is the management system verification phase. This phase defines and measures indicators to assess the progress of the implementation of the action plan and the effectiveness of the measures deployed. This phase can be carried out:

– as part of a self-checking process;

– via supervision indicators;

– through internal audits of the organization;

– via penetration tests.

This step can be seen as the implementation of a security dashboard.

Phase 4 is the Act phase, the improvement of the industrial security management system. The objective of this phase is to improve the security management system. A first part concerns the treatment of non-conformities detected by the previous phase. A second area concerns improvement of the ISMS itself to ensure that it remains relevant.

3.7. Risk treatment process

Risk management is central to this approach. It is formalized in the ISO 27005 standard by the risk treatment process. This process includes the risk analysis and treatment steps described in section 3.4.

It builds on the initial step of defining the ISMS context and provides a communication step throughout the process. This communication includes, on the one hand, consultation with stakeholders to carry out a relevant risk analysis and, on the other hand, informing the entities concerned about the consequences of the risk. Finally, the risk treatment process must be monitored and improved; this is the purpose of the monitoring and review phase, carried out as part of the Check phase of the ISMS.

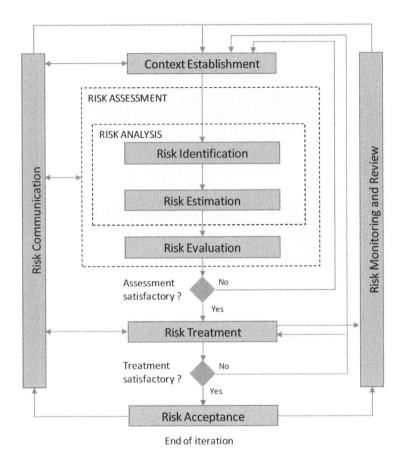

Figure 3.12. *Risk management process (27005)*

3.8. Governance and security policy for IT systems

3.8.1. *Governance*

For the risk management system to work, particularly with regard to the continuous improvement approach, a strong commitment from senior management is required, first to take the decision to implement a risk management system and then to define a strategic vision of security that is reflected in the development of a security policy. This policy is published and then implemented in an operational manner.

Management support must also be provided at hierarchical level to ensure that the approach is implemented by the entire organization, and at the budgetary level, to give stakeholders the means to apply and sustain the security system.

Within the framework of the 27001 standard, paragraph 5, leadership, formalizes this step of management engagement.

3.8.2. Security policy

An Information Systems Security Policy (ISSP) expresses the management's desire to protect the organization's information and associated technological resources. It is essential for security governance.

It defines the security objectives that have been expressed in the risk treatment step, or provides a framework for defining them. It includes a commitment to meet applicable security needs and to achieve continuous improvement.

In general, the ISSP should address the following points:

– the importance of protecting and building trust: determining issues for the company, the regulatory aspects to be considered, the threats to be taken into account;

– what needs to be protected and at what level, resulting in a list of assets (tangible and intangible) to be protected and a scale of needs;

– who is responsible for protection, which is defined in terms of organization, responsibilities and management of the ISS;

– how to protect the different elements, resulting in a coherent and operational set of security rules;

– when to protect, taking into account the entire lifecycle.

An ISSP may, for example, address the following aspects:

– context and objectives: the company, the scope of the ISS, security needs;

– organization, governance and accountability;

– human resources;

– asset management;

– integration of ISS into the lifecycle;

– physical security;

– access control;

– taking into account the IS in operation of the information system;

– workstation security;

– network security;

– development security;

– third-party management;

– incident handling;

– business continuity;

– compliance, audit and controls.

The prior completion of an ISS risk analysis facilitates the development of an ISSP, particularly to determine the context and scope, select security objectives and guide rule making, and ensure consistency with the security needs identified for the organization. An example of an ISSP is the French State Information Systems Security Policy (ANSSI 2014c).

The ISSP is developed within the framework of a project organization: a designated project manager, one or more working groups are formed, resources are allocated, a timetable is defined and deliverables are identified.

3.9. Security management of industrial systems

The management of risks related to industrial cybersecurity is similar to that of IS risk management and risk management in general: management must set objectives adapted to the challenges and define a strategy. A phase prior to the security of an industrial IS is therefore to make management aware of the stakes of the cybersecurity issues and make management aware of the company's vulnerabilities.

The ICS security policy must be integrated with the existing IT security policy. Very often, the company's organization does not promote the links between the world of traditional information technology and that of industrial systems.

The development of the IT security policy for industrial systems (I-ISSP) must be co-constructed with the personnel in charge of operations. This can be done in three steps:

1) Create a working group involving the operational referent staff (operation and maintenance manager, automation engineers, industrial IT manager, industrial IS architect, etc.) to carry out a more or less detailed risk analysis (Chapter 9).

2) Structure the results by adopting a graduated approach based on the level of risk per area (Chapter 7), and combine them with appropriate security measures. Monitoring indicators can be defined at this stage.

3) Develop an I-ISSP application framework that takes into account the specific zoning and lifecycle of industrial information systems.

This I-ISSP will then be implemented at the operational level by appointing a functional chain, and by setting up a project-type organization, with a deadline (a few years maximum), allocated resources and relevant indicators. It may be interesting to start with the most at-risk areas and define quick wins, such as inventory coverage rate, removable media usage rate, open flows and awareness actions.

4

Threats and Attacks to ICS

4.1. General principle of an attack

In general, an industrial control system (ICS) is subject to threats, generated by sources of threats. These sources of threats use an attack vector to carry out an attack.

The attack may be directed at the control command system (BPCS) that controls the physical system, or it may be directed at the safety instrumented system (SIS), so as to create damage when the system uses safety functions.

The path or means used to access the target system is called the "attack vector".

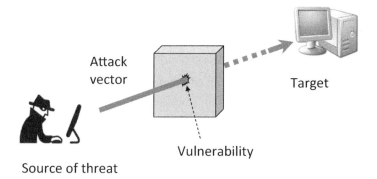

Figure 4.1. *Vector of attack and vulnerability. For a color version of this figure, see www.iste.co.uk/flaus/cybersecurity.zip*

The objective of a computer attack is to create damage to one of the installation's assets. Different types of damage are possible:

– the asset may be corrupted, so that the system performs incorrect actions, or provides incorrect results, or information is lost;

– the asset can be spied on: information related to the asset is accessible to persons who should not have access to it;

– the property may become unavailable or very slow, so that use of the system or network becomes impossible or impractical, which is particularly problematic for a real-time control system or a security system.

To successfully create this damage, an attacker has a number of options that may be logical or physical, i.e. through computer access or physical access.

Logically, an attacker may attempt to access the content of the memory of the attacked device, to copy or modify it, or will attempt to have the device execute a series of commands. It may also seek to spy on or modify communications between the device and its environment.

Recent devices use identification and authentication mechanisms coupled with rights control, which means that, to execute commands or access memory, a user, who may be a human or a machine, must prove its identity. When this is done, they are given a number of options to read or write data, or to execute commands. The older devices do not have these possibilities.

Similarly, communications between two devices may use identification and authentication mechanisms, and in some cases some protocols be encrypted, i.e. made unreadable by an observer who does not have the mechanism or code for decryption. Again, these possibilities only exist for recent equipments.

To slow down a system, an attacker has two main possibilities:

– if he/she has successfully entered a system, he/she can seek to initiate the execution of many processes;

– from the outside, he/she can flood the system with requests to prevent its normal operation, as in the case of a denial-of-service (DoS) attack by machine networks that overload a server. He/she can also, if he/she has access to the network, send it restart or resynchronization commands, which will disrupt its operation.

To these possibilities are added the physical actions performed on the device. An attacker can:

– steal the device;

– recover it (management of discarded equipment, loss of keys, etc.);

– add a spy device (keylogger, for example);

– destroy it.

At this stage, we can distinguish several types of actions that have a negative impact on the information system:

– unintentional actions, such as a user mistakenly formatting a storage device;

– malicious actions by users with authorized access to the system, such as an administrator from whom a company has separated and who wishes to harm it;

– malicious actions by unauthorized persons who have successfully obtained legal identifiers (fraud, deception, brute force, use of default identifiers, etc.);

– malicious actions without identifiers, which use different techniques to bypass identification and authentication mechanisms.

Only the last category is based on technical skills and hardware and software vulnerabilities; the first three are related to the human factor and organizational aspects.

From a technical point of view, in the latter case, an attacker will attempt to bypass the identification and authentication system, on hosts or in network exchanges, and/or attempt to bypass the rights management system to successfully perform actions beyond what is allowed. To do this, different techniques are used and are presented below.

It is important to note that, in the case of ICS, there are still many devices for which identification is not required, and that most protocols are not secure. If an attacker succeeds in accessing the industrial network, he/she can relatively easily listen and modify exchanges, and he/she can act on the connected devices.

Different means can be combined to penetrate a target system and develop the ability to perform malicious actions in a sustainable manner. This is called an advanced persistent threat (APT) attack. It goes through several stages:

– a recognition phase: research, identification and selection of targets and their vulnerabilities. The targets can be technical (servers, PLCs) or human (personnel in the target company);

– a phase where the malware (often of the backdoor type) is transmitted to the victim by email, via a malicious website or removable memory, or by exploiting a vulnerability in the target system;

– an exploitation phase: once the malware is installed secretly on the target, it spies, or waits for commands from the attacker. It can also attempt to copy itself onto the other stations of the network.

Many ICS attacks exploit this principle, such as the Stuxnet attack or the attack on the Ukrainian electricity distribution system presented below.

A primary attack can also generate a secondary attack (Figure 4.2), the objective of the primary attack being to install malware that can be used as a relay for other attacks (Hutchins *et al.* 2011).

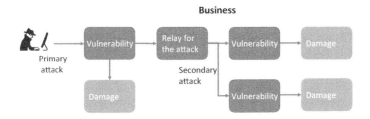

Figure 4.2. *Primary and secondary attacks. For a color version of this figure, see www.iste.co.uk/flaus/cybersecurity.zip*

A complete threat assessment for an information system reveals threats other than those linked to computer attacks (Lévy-Bencheton *et al.* 2015). We can mention the threats:

– linked to physical phenomena such as fire, floods, etc.;

– linked to physical sabotage, which can impact communications in particular;

– due to equipment failures;

– due to energy losses for operation and cooling;

– due to human factors: pandemic, strike, various absences, etc.

These threats do not fall into the category of attacks, but are to be taken into account in the risk analysis.

4.2. Sources of threats

As presented in the previous section, there are several categories of threat sources. A classification is useful for risk analysis because, on the one hand, it allows all relevant sources to be systematically considered and, on the other hand, it makes it possible to choose countermeasures that must be adapted to the source's capabilities.

An initial classification is provided by the NIST guide SP800-82. It shows all sources capable of causing damage to an information system (Table 4.1). It can be noted that most of them are not of human origin.

For those of human origin, in the National Institute of Standards and Technology (NIST) classification or in that proposed in the EBIOS method (Table 4.2), a distinction is made between violations that are caused by human error without malicious intent and those that are malicious. It is the latter that are taken into account, explicitly or not, when considering cybersecurity.

It is useful to distinguish between internal and external threats, as they operate in a completely different context. Internal threats have authorized access, based on their access rights, and can therefore cause damage without the need to pass a technical vulnerability. If they do, they can operate from the local network, and therefore have extensive possibilities. Typical examples are those of a former malicious employee or a malicious external provider. Countermeasures for internal threats are essentially organizational: limited access rights, double validation for administrators, management of staff arrivals and departures, etc.

External threats must use other vectors of attack, such as the Internet, or social engineering. In order to implement appropriate countermeasures, the following levels are distinguished:

– malicious people using robots that systematically and automatically "scan" for vulnerabilities;

– amateur hackers, with limited capabilities, who find attack software on the Internet, and who do not necessarily target a given installation;

– advanced hackers, with advanced technical skills, such as pirate groups;

– organizations with very large or even unlimited resources, such as States, criminal organizations or, to a lesser extent, a competing company.

This classification is used in the context of risk analysis. Depending on the installation under consideration, risks for the relevant categories are analyzed. The company must ask itself who can attack it and why.

A more detailed catalogue of sources of threats may be useful. They are associated with the various risk analysis methods (Chapter 9) or in the NIST SP800-82 standards (Stouffer*et al.* 2015), NIST SP 800-30 (Joint Task Force 2012), ISF IRAM (Jenkins 2014), ISO 27005 (IEC 2011) and BSI Threat Catalog (Bundesamt für Sicherheit 2011).

Type of source	Description
Malicious human action	

– Individuals
   - External
   - Internal
   - Trusted internal
   - Internal with privilege
– Groups
   - *Ad hoc*
   - Established
– Organizations
   - Competitor
   - Supplier
   - Partner
   - Customer
   - State or nation
	Individuals, groups, organizations or States seeking to exploit the organization's dependence on information resources (for example, information in electronic form, information and communication technologies, as well as communications and information processing capabilities provided by these technologies)

Human error	
– User – Preferred user/administrator	Unintentional actions carried out by individuals in the course of carrying out their daily responsibilities
Technical system	

– Information technology equipment
   - Storage
   - Processor
   - Display system
	Equipment failures, lack of environmental control, software design or implementation defects, resource depletion

Type of source	Description
Technical system	

- Sensor
- Controller
– Utilities
  - Temperature/humidity control
  - Power supply
– Software
  - Operating system
  - Network
  - General applications
  - Specific applications	

Natural and anthropogenic environment	

– Natural or human catastrophe
  - Fire
  - Flood/tsunami
  - Storms/tornadoes
  - Earthquake
  - Bombing
  - Overflow
  - Unusual natural event
– Infrastructure failure
  - Telecommunications
  - Electrical power
	Natural disasters and failures of critical infrastructure on which the organization depends, but which are beyond the organization's control

Table 4.1. *Different types of sources (adapted from NIST SP800-82)*

Deliberate human sources
Internal human source, malicious, with low ressources
Internal, malicious human source, with significant resources
Internal human source, malicious, with unlimited resources
External human source, malicious, with low resources
External, malicious human source, with significant resources
External, malicious human source, with unlimited resources
Human sources acting in an accidental manner
Internal human source, without malicious intent, with low ressources

Internal human source, without malicious intent, with significant capabilities
Internal human source, without malicious intent, with unlimited capabilities
External human source, without malicious intent, with low ressources
External human source, without malicious intent, with significant capabilities
External human source, without malicious intent, with unlimited capabilities
Non-human sources
Malicious code of unknown origin
Natural phenomenon
Natural or medical disaster
Animal activity
Internal event

Table 4.2. *Source classification (EBIOS)*

4.3. Attack vectors

To carry out an attack, a threat source uses a means or vector of attack. Table 4.3 presents the main vectors used in ICS. The choice of attack vector is made in such a way as to exploit a system vulnerability.

Category	Threats and vectors of attack
Network	Adjacent internal wired networks such as IT network or DMZ using compromised equipment
	Compromised equipment with multiple network interfaces and belonging to two networks
	Wi-Fi networks
	Compromised equipment on the local network
	Internet connections
	Cloud connections
	Split tunneling (simultaneous access to protected resources via VPN and Internet for example)
	Wireless Wi-Fi networks
	Industrial or IoT wireless networks (Hart, Zigbee, etc.)

Physical access	Removal media on USB port or other USB keyboard socket (keylogger) Serial and other ports (SATA, Display port, HDMI, etc.) RJ45 ports Direct access via keyboard/mouse Flash memory cards
ICS equipment	USB and other ports Open network logic ports (http, ftp, etc.) Configuration or program transfer software Traffic sent to the equipment (Modbus, etc.)
Applications	Network logic ports User data entry (local or Web interface) Data entry files Data reading via libraries OS vulnerability (e.g. data access via other applications)
People	Social engineering by email Social engineering by telephone Corrupted client email and attachments Internet browser
Suppliers	Supply of electronic components, boards Updates or supply applications, operating system or firmware

Table 4.3. *Main vectors of attack*

4.4. Main categories of malware

A computer, network equipment, ICS component or IoT equipment are programmable devices with a memory area containing both data and programs. Some zones are not modifiable, some others are modifiable, but keep the data when the power supply is interrupted, some still are cleared when the device is switched off. When a system is started, it executes a predefined and unmodifiable program that allows us to start a start program located at a given address. An attacker can try to modify the memory areas, insert malicious code into the boot address, or execute this malicious code by modifying the stack (section 4.5.1).

In addition, systems connected to a Transmission Control Protocol (TCP)/Internet Protocol (IP) network operate by listening to the network via logical ports. We talk about "open ports". Each port is assigned a default software: for example, port 23 is listened to by Telnet and port 502 is used by Modbus. An attacker can therefore try to use vulnerabilities related to protocols and applications using these ports to modify data, steal passwords or execute code on a workstation or equipment.

These various possibilities are exploited by malware. We distinguish different categories presented in the following sections.

4.4.1. *Virus/worms*

A virus is malicious software that installs itself in a hidden way on a computer or equipment and is able to replicate itself from one computer to another. In general, to infect a system, a virus attaches itself to another software. When the latter is executed, the virus is executed and can infect the host machine and attempt to replicate itself. The biological metaphor is adapted to describe the behavior of this type of program. However, to contaminate or replicate, it must be executed via the contaminated software or directly. This can happen automatically, in the case of USB sticks with a self-executing file. Most often, it is the user who runs the program himself, after being misled about the nature of some software or of an attachment.

A worm is a type of virus that spreads over the network, often using a vulnerability to perform remote execution *over* the network. Nowadays, there is no longer any real distinction between worms and viruses.

4.4.2. *Trojan horse*

A Trojan horse is software that seems to do one thing, and can, in fact, do it, but embeds malicious software. The analogy to the famous story from Greek mythology is very appropriate: the giant wooden horse, an offering in appearance, actually hid Greek warriors. In the case of computers, the Trojan horse deceives the user and makes sure that the hidden malware is installed on his/her machine. The latter can then perform its hidden actions, for example spying and sending the user's information to a pirate server.

4.4.3. *Logical bomb*

A logical bomb is a category of malware that installs itself in a hidden way, in a Trojan horse for example, and triggers actions at a date predefined by the hackers.

4.4.4. *Rootkit*

A rootkit is malicious software that installs itself in a hidden way, and whose purpose is to provide privileged user access to an attacker, who therefore has almost all the privileges to perform all actions on an operating system.

4.4.5. *Spyware*

Spyware is software that spies on users and records their activities. As a general rule, it is installed without the user's knowledge. Spyware can perform a wide range of activities. It can record keystrokes (an action commonly known as keylogging), when the user connects to specific websites. It can record sound and/or image or anything else.

It can use a backdoor to send its information to a hacker server.

4.4.6. *Back doors*

A backdoor is a hidden feature of a software or operating system that is most often introduced at the time of development and allows access to specific options or bypassing the normal authentication procedure. Originally, the objective was to allow the recovery of lost passwords or simply for entertainment purposes, as was the case with software that displays special animations in the *About* window when typing a special sequence. With the networking of computer systems, this functionality has become a vulnerability, especially when authentication is hard coded.

In addition, there are many cases where the back door is implanted during manufacture in software, equipment (camera) or telephones for illegal purposes, such as data theft. There are also many malware programs that implement such a feature.

4.4.7. Botnet

A botnet is a network of machines connected to the Internet, each of which executes a "bot", i.e. robot software responsible for responding to commands or executing predefined actions. Malicious "robot" software is distributed by one of the means mentioned above, such as viruses or emails. They are installed silently on targeted workstations or equipment and wait for orders. The consequences are benign for the user on the surface, however, this makes it possible to build armies of machines equipped with malicious software, the botnet.

These zombie machines, which can be computers or equipment connected to the Internet such as cameras or industrial Internet of Things (IIoTs), are used to carry out other attacks. They can send spam or perform distributed denial of service (DDoS) attacks.

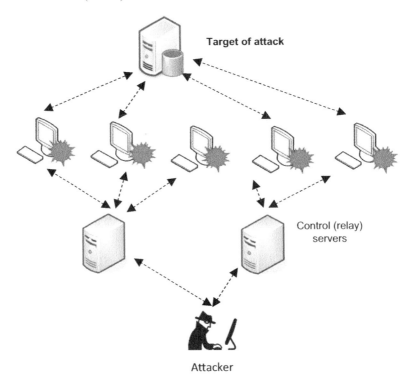

Figure 4.3. *DDoS attack by a botnet. For a color version of this figure, see www.iste.co.uk/flaus/cybersecurity.zip*

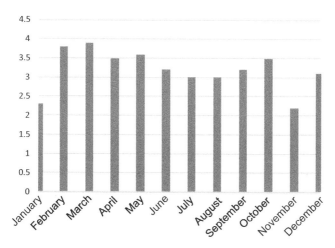

Figure 4.4. *Percentage of ICS computers attacked by a botnet in 2017 (source: CERT)*

4.4.8. *Ransomware*

Ransomware is a type of malware that prevents or limits users' access to their system, either by locking the system screen or by locking users' files. To recover access, a ransom must be paid. The most modern ransomware families, collectively classified as crypto-ransomware, encrypt certain types of files on infected systems and force users to pay the ransom via certain anonymous online payment methods to obtain a decryption key (cryptocurrency).

4.5. Attacks on equipment and applications

4.5.1. *Buffer overflow and integer overflow*

A buffer overflow attack attempts to place more data in a memory area than it can contain. This can corrupt other variables, block the system, or, when the zone is properly placed, write to the subroutine call stack and cause malicious code to be executed. In Figure 4.5, the variable buffer is a table of four characters, numbered 0 to 3. Writing beyond that is illegal. If no control is performed, writing from buffer (Abshier 2004) to buffer (Pollet 2010) allows to modify the return address, and thus cause the execution of a malicious code that will have previously been loaded at the chosen address. More details can be found in Howard and Leblanc (2002).

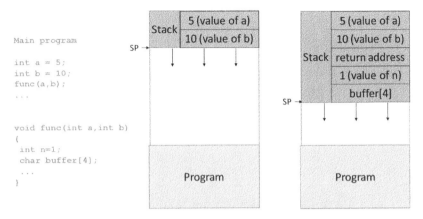

Figure 4.5. *Attack by buffer overflow. For a color version of this figure, see www.iste.co.uk/flaus/cybersecurity.zip*

There are also vulnerabilities related to integer overflow. These vulnerabilities are often found by random software testing (fuzzing tests). A flaw of this well-known type is related to the Secure SHell (SSH). By sending a frame designed to exploit the flaw, was possible to get connected as "root" (CERT VU#945216 vulnerability).

4.5.2. *Attack by brute force*

A brute force attack is a trial and error method used by some software to decode encrypted data such as passwords or data encryption keys. Just as a criminal can "break" a safe by trying many possible combinations, a brute force cracking application goes sequentially through all possible combinations of legal characters. Brute force is considered an infallible approach, although it takes a long time. In reality, software programs performing this type of attack combine brute force and a dictionary approach: they will first test the usual passwords, then the dictionary words, then classical variations, before systematically testing all combinations.

Encryption systems are designed so that the time required is such that this method is not feasible on a human scale. However, new means of calculation (quantum computing) could reduce this time. For this reason, some guides require that the authentication system can be updated on equipment, such as the Industrial Internet Consortium's IIoT good practices described in Chapter 6.

This approach cannot be used directly, as most systems crash after a few attempts. This approach is most often used on encrypted password files that have been previously stolen.

Finally, it should be noted that the effectiveness of this type of attack is closely linked to the strength of the passwords, which must be chosen in a relevant way (ANSSI 2012b).

4.5.3. *Attack via a zero day flaw*

A "zero day" attack is an attack that exploits vulnerabilities that have not been fixed or made public. Often, this type of attack includes attacks targeting vulnerabilities known to the public but not yet fixed.

Software vulnerabilities can be discovered by hackers, security companies, researchers, government intelligence services, software vendors themselves or users. If discovered by hackers, they can be used to set up attacks also called "exploits". An exploit will be kept secret for as long as possible. It can be sold on the cybercrime black market.

4.5.4. *Side-channel attacks*

In general, side-channel attacks use information leaks such as execution time, power consumption or electromagnetic leaks observed during the normal execution of a cryptographic algorithm to infer secret information such as encryption keys (Appendix 1).

It is important to note that the quality of the algorithm is not in question, either its principle or its implementation, rather it is observation of its functioning that allows it to be bypassed. In a way, this method is similar to opening a safe with a stethoscope.

The most well-known approaches (Joye and Olivier 2011) use:

– consumption analysis, either directly during an encryption operation (single power analysis) or by making repeated observations and using statistical methods (differential power analysis);

– analysis of computing time (Kocher 1996), for example to validate or invalidate a password;

– the electromagnetic waves emitted.

The risk associated with this type of attack depends on the physical accessibility of the equipment, which can be easy for an Internet of Things (IoT) device, and the level of integration of the component. It is more complex with a System on a Chip (SoC, section 11.12.4), and countermeasures to smooth consumption are implemented in recent circuits.

4.5.5. *Attacks specific to ICS equipment*

Equipment such as PLCs are particular computer systems, which can therefore be subject to the same attacks. For example, since many PLCs work with VxWorks, they have vulnerabilities such as the CVE-2015-7599 vulnerability. This, due to a problem of unchecked integers overflow, allows a user to execute arbitrary code or to create a DoS.

However, ICS equipment also has functional vulnerabilities: it is possible, by using "normal" commands in an inappropriate way, to create malfunctions. These commands are, for example, a reset device, sending a malformed request, forcing a stop, synchronizing the clock, etc. It is therefore important to prevent any intrusion into the industrial network in order to avoid the generation of commands of this type.

It is also possible to insert malware into the program sent to the PLC (Govil *et al.* 2018). This program is written in one of the languages specific to PLCs (Chapter 1), such as the Ladder Language. However, it is possible to insert malicious actions that affect the actions, the measurement values or even create a DoS. Transfers to the PLC are not very secure, so this type of attack should be considered.

Other attacks to consider are those related to the modification of the PLC firmware (Basnight *et al.* 2013; Schuett *et al.* 2014).

Another type of attack concerns the OPC/DCOM protocol. The OLE for Process Control (OPC) version is a technique that appeared in 1995 and is not very secure (Chapter 2). OPC allows an attacker to list system properties or exploit vulnerabilities by buffer overflow. The OPC protocol has been replaced by the OPC/UA standard proposed by the OPC Foundation, which is much more secure.

Dragonfly malware has used these vulnerabilities. The Havex module uses the OPC standard to collect information on industrial control devices. It then sends them to the command and control server (C&C) where they are analyzed by hackers.

4.5.6. Attacks on IIoT systems

IIoT-based systems are networked systems in which each node is a more or less sophisticated programmable device. It is subject to the same vulnerabilities as traditional computer equipment.

In addition to traditional attacks, IIoT-based systems are sensitive to specific attacks and can be used as a platform to conduct specific attacks.

Among these specific attacks, we can mention the following:

– physical attacks, by auxiliary channel, by observing consumption to deduce the keys (section 4.5.4) (Mangard *et al.* 2007) or to carry out reverse engineering;

– attacks on energy resources, by preventing the equipment from going into rest mode to exhaust its energy (denial of sleep) (Raymond *et al.* 2009);

– attacks on system software, which are complex to update;

– attacks on network addition and removal features.

These aspects are detailed in Chapter 5, on vulnerabilities.

It should also be noted that the stakes of data attacks may be higher than in the case of traditional ICS. Let us quote, for example:

– attacks aimed at recovering data to build an image of the activity of a company, a person or a manufacturing process (for example, deducting that a place is occupied from the consumption or switching on of lamps);

– attacks aimed at modifying data transmitted in order to fraud, such as Spanish electricity meters (Illera and Vidal 2014).

Conversely, IIoT equipment can be used as an attack platform, and because of their large number, they present a significant threat. This is the case when:

– they are corrupt, to enroll them in a botnet in order to carry out DoS attacks;

– they are used to saturate the local communication network (radio or Wi-Fi).

In more traditional attacks on IoT devices, malware attempts to modify the firmware, or insert malicious code at boot time. IIoTs are more vulnerable because they are often physically isolated and can be the subject of electronic analyses or reverse engineering.

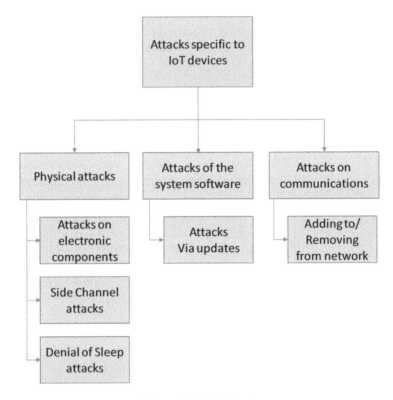

Figure 4.6. *IoT attacks*

4.6. Site attacks and via websites

The Structured Query Language (SQL) injection attack is an attack in which malicious code is inserted into strings that are then transmitted to an SQL server instance for analysis and execution. Any procedure that builds SQL statements must be checked for injection vulnerabilities, as an SQL server will execute all syntactically valid queries it receives. Even the queries with parameters can be manipulated by a qualified and determined attacker.

A simple example is the following instruction, in which the grayed-out term has been added:

SELECT * FROM Users WHERE UserId = 105 OR 1=1;

As this term is always true, it makes it possible to select all users. This type of attack is particularly dangerous when used on a website.

A Cross-Site Scripting (XSS) attack is a type of attack in which an attacker injects data, such as a malicious script, into content from trusted websites. Cross-site script attacks occur when an unreliable source is allowed to inject its own code into a web application, and this malicious code is included in the dynamic content provided to the victim's browser. This is the case, for example, in forums, if the content entered by users is not validated. The injected code is often Javascript, and can be used to steal identification data. More details are provided in (DGS 2007). An example of a possible attack on an industrial system is explained in the ICS-ALERT-13-304-01 vulnerability. This flaw affected the SCADA/HMI Nordex Control 2.

4.7. Network attacks

4.7.1. *Man-in-the-middle*

Figure 4.7. *MitM attack. For a color version of this figure, see www.iste.co.uk/flaus/cybersecurity.zip*

The principle of Man-in-the-Middle attacks (MiM or MitM attacks) is to pass communications between two stations through a relay without the knowledge of the two stations in communication (Figure 4.7). This can be done using different techniques:

– the poisoning of the ARP cache (ARP Cache Poisoning) described in Chapter 2: if both workstations are on the same local network, it is possible, even relatively easy, for the attacker to force the communications to pass through his computer by posing as a "relay" (router, gateway). It is then quite simple to modify these communications;

– DNS-Spoofing: a DNS server translates a site name (for example www.myscadasupplier.com) into an IP address. The attacker alters the DNS server(s) in order to redirect to it communications for a website;

– Internet Control Message Protocol (ICMP) redirection: using the ICMP, an attacker can send a fake message to a router to redirect a victim's data flow to his/her own machine. This option must therefore be blocked in routers.

With this type of attack, an attacker cannot only capture all traffic, including sensitive data such as usernames and passwords, but can also delete connections at will and manipulate content to deceive the victim. This attack works even if the traffic is encrypted, as the attacker can substitute his/her private/public keys when establishing the call.

4.7.2. Denial of service

A DoS attack is an attack designed to prevent a system or service from operating normally. It can exploit a known vulnerability in a specific application or operating system, or use certain vulnerabilities in specific protocols or services. In a DoS attack, the attacker tries to prevent authorized users from accessing either specific information or the computer system or the network itself. This can be accomplished by causing the system to shut down unexpectedly (e.g. by buffer overflow) to cause a cessation of service or by flooding it with requests (e.g. with a botnet).

A DoS attack can also be used in conjunction with other actions to gain unauthorized access to a computer using a MitM attack.

In the context of ICS, a DoS of a system in charge of controlling a physical equipment can have serious consequences, since the physical system continues to evolve over time (it is said to be real time). If the system under attack is the SIS, the security of persons and property may be affected.

A DoS attack conducted by a network of computers or connected objects is called a DDoS (Distributed DoS) attack. It can also be conducted using a TCP/IP vulnerability (SYN flood) (Chapter 2).

4.7.3. Network and port scanning

Computers and equipment connected to a TCP/IP network communicate using ports. There are 65535 TCP and UDP ports that can be opened, i.e. for which the system is ready to open a communication. A number of ports from 1 to 1,023 have a predefined role: for example, port 21 is used for FTP communications, port 80 for the web server. Some of the communication modules ready to accept a communication on one of these ports have vulnerabilities: for example, FTP servers on port 21 may allow anonymous connections. These are of course ideal targets and significant vulnerabilities (Mathew et al. 2014).

During an intrusion attempt, a port scan allows us to know if there is a device at an address, and then to know its functionalities and possible vulnerabilities. The approach of the analysis is often as follows:

– test to see if the host is active;

– test to know if the host is behind a firewall;

– detection of the OS or the type of equipment;

– port scanning;

– looking for vulnerabilities.

The intrusion attempt is performed by testing known vulnerabilities on identified ports. For example, for port 80, which is the port used for websites, automatic software sends requests to attempt SQL injections, cross-site scripting attacks or buffer overflow attempts. For port 22, used by SSH, connection attempts can be made by a brute force attack.

The nmap software allows "scanning" a computer network and opens ports. The PLCSCAN utility allows detection of the PLCs present on a network (Figure 4.8).

Detecting port scanning activity is therefore important, as it is a precursor incident that can identify an attempted attack.

Usage examples

```
plcscan.py --timeout 2 192.168.0.1:102 10.0.0.0/24
```

Output examples

```
127.0.0.1:102 S7comm (src_tsap=0x100, dst_tsap=0x102)
Module : 6ES7 151-8AB01-0AB0 v.0.2 (36455337203135312d38414230312d304142302000c000020001)
Basic Hardware : 6ES7 151-8AB01-0AB0 v.0.2 (36455337203135312d38414230312d304142302000c000020001)
Basic Firmware : v.3.2.6 (20202020202020202020202020202020202020200c056030206)
Unknown (129) : Boot Loader A (426f6f74204c6f61646572202020202020202020000041200909)
Name of the PLC : SIMATIC 300(xxxxxxxxx) (53494d4154494320333030280000000000000000002900000000000000000000).
Name of the module : IM151-8 PN/DP CPU (494d3135312d3820504e2f4450204350550000000000000000000000000000)
Plant identification : (0000000000000000000000000000000000000000000000000000000000000000)
Copyright : Original Siemens Equipment (4f726967696e616c205369656d656e732045717569706d656e74000000000000)
Serial number of module : S C-B0UVxxxxxxxx
(5320432d424f5556xxxxxxxxx00000000000000000000000000000000)
Module type name : IM151-8 PN/DP CPU (494d3135312d3820504e2f4450204350550000000000000000000000000000
```

Figure 4.8. *Network scanning*

4.7.4. *Replay attack*

A replay attack is an attack that consists of intercepting packets from the network, possibly by a MiM mechanism, and then reusing them to forward them to the recipient. It is not necessary to understand or decipher them.

For example, if a sender sends an encrypted name and password to a recipient to authenticate himself, it is possible for a hacker, by reusing the same sequence, to usurp the sender's rights.

An attack of this type is possible against certain types of vulnerable PLCs: an attacker with access to the OT network can steal session numbers and add arbitrary commands to stop or restart the PLC.

4.8. Physical attacks

If the attacker can obtain physical access to the target, a number of additional attacks are possible.

First, it can simply destroy or degrade the equipment. This possibility aside, he can:

– connect devices to spy on data (including passwords), such as a keylogger that connects to the keyboard wire and captures all user entries, or via the JTAG test port;

– steal memory cards or mass memories to copy or modify them;

– steal equipment in order to carry out reverse engineering and possibly detect vulnerabilities that can be used remotely;

– simply use the live keyboard-mouse device to attempt an intrusion or the screen to steal information.

A physical attack may also be carried out on discarded equipment for the purpose of stealing data or passwords.

4.9. Attacks using the human factor

4.9.1. *Social engineering*

One of the weakest links in IT system security is the human aspect. In addition to human errors, unintentional (carelessness) or deliberate (conscious non-compliance with the rules), the human user can be a victim of so-called social engineering techniques.

These methods are based on lies, misrepresentation, blackmail or greed. The attacker can, for example, contact a system administrator and pretend to be an authorized user, asking to have a new password. Another common strategy is to pose as a member of a supplier's staff who needs temporary access to perform emergency maintenance.

These techniques also rely heavily on phishing scams that aim to abuse the "naivety" of users to retrieve their credentials.

There are two types of phishing: mass phishing, using generic emails, and spear phishing, which is carried out after investigation of the target and the company, and can be much more difficult to detect.

Examples of an attack may include the following:

– the receipt of an email using the company's logo and colors;

– a request to perform an operation such as updating personal data or confirming the password;

– a connection to a fake site identical to that of the company and controlled by the attacker;

– the recovery by the attacker of the login/passwords (or any other sensitive data) entered by the customer on a fake site.

This site may be an equipment manufacturer site and, for example, provide updates.

Such an attack occurred in early 2018: a fake patch for Meltdown/Specter Patch vulnerabilities attempted to install malware. The emails appeared to come from the German Federal Office for Information Security (BSI), and linked to an imitation site, with an SSL certificate (https:). The site was not official, but tried to encourage users to install the patches, which were actually malware.

4.9.2. Internal fraud

The fact that an authorized user knowingly uses his or her access to harm the security of an organization is called internal fraud. The user may be an employee, a consultant or a subcontractor. This case is to be distinguished from the user who provides their access details, either through negligence or because they has been misled. This problem remains a taboo subject for many companies, while it is increasing significantly (Cole 2017). There are different categories of fraudsters: the occasional fraudster, the recurrent fraudster, the person who is deliberately hired to carry out fraud and group fraud. The problem is particularly important for privileged users such as managers or administrators (Ware *et al.* 2017).

These attacks are made possible by organizational vulnerabilities:

– weaknesses in internal control and operational monitoring procedures;

– permissive management of IT authorizations;

– lack of segregation of duties and rotation.

Appropriate organizational measures are therefore to be planned.

4.10. History of attacks on ICS

The first known attack on an ICS was in 1982 on a gas pipeline in Russia. A Trojan horse introduced into a Supervisory Control And Data Acquisition (SCADA) system caused the pipeline to explode. Since then, many attacks have taken place. Table 4.4 presents the most well-known attacks.

An important milestone was the 2007 attack demonstration by the Idaho National Laboratory. This one, called Aurora, was done to demonstrate how a cyber-attack could destroy the physical components of the power grid. In this attack, the attacker uses a communication protocol vulnerability to access the control network of a diesel generator. This allows it to run a malicious computer program that has been designed to open a circuit breaker, wait for the generator to desynchronize and immediately close the circuit breaker. Such actions, carried out fast enough so that the protection system cannot detect the problem, led the diesel generator to explode, which can be seen on the video of the experiment (Idaho National Laboratory n.d.). Since most electrical network equipment uses this type of communication protocol, the experience highlights a vulnerability that is of particular concern.

A few years later, Stuxnet (Falliere *et al.* 2011) made its mark on the history of cybersecurity of industrial systems and raised awareness of the vulnerability of SCADA systems. According to anonymous sources, Stuxnet was a computer worm developed to slow down Iran's nuclear program. Stuxnet specifically targeted PLCs used by centrifuges to separate nuclear materials for uranium enrichment purposes. It would have allowed the destruction of 20% of production capacity.

The principle of Stuxnet's operation is, on the one hand, to propagate between the machines of a Windows network, and on the other hand, when it detects a programming environment for a PLC of a certain type, to set up a module that modifies the program sent to the PLC when it is transmitted. This modification alters the behavior of the control system and degrades the centrifuges controlled by untimely and hidden speed variations.

In more detail, Stuxnet is introduced into the target environment via a drive connected to a USB port. The worm then spreads across the computer network using several Windows vulnerabilities:

– it replicates itself via removable drives exploiting a vulnerability allowing automatic execution via LNK/PIF files (BID 41732);

– it spreads in a local network via a vulnerability in the Windows spooler (print queue manager), which allows remote code execution (BID 43073);

– it spreads via SMB, a shared file access system, by exploiting the Remote Procedure Call (RPC) remote code execution of the Microsoft Windows Server service (BID 31874).

Once installed on a workstation, if this workstation runs the Step 7 development environment, Stuxnet modifies the library that communicates with the PLC. It can therefore add the malicious code to the sent program and hide the return from the PLC.

Another turning point occurred very recently in 2017. This is the attack called TRITON (Ferguson 2018), designed to attack Triconex security systems, particularly those installed in industrial petroleum units. The precise location of the facilities affected by the attack has not been revealed, but it is assumed that these units were located in Saudi Arabia. The objective of this attack was to disable the security systems. From this situation, it is possible either to shut down the unit using the SIS or to allow the installation to pass through in a dangerous state. In the case of a process control error, either by accident or caused by another attack, the consequences could have been dramatic. A bug discovered the attack.

TRITON's intention to harm is clear, since this type of system is intended to guarantee the physical safety of the facility and the population. Causing a SIS to malfunction means ensuring that an industrial accident can occur in the event of operational disruptions.

At the IoT level, a well-known demonstration is the attack on Philips Hue lamps in 2016. This demonstration was carried out by researchers (Ronen *et al.* 2017) who succeeded in taking control of a network of connected Philips Hue brand lamps. These bulbs are connected via a Zigbee network, and the user can use a remote control to control the intensity or color of the light they emit. These bulbs had a significant security breach.

As the researchers explain, the encryption keys used by the connected bulbs to communicate with each other are all the same, allowing a bulb infected with malware to easily transmit it to other lamps. The flaw exploited by the researchers is the Zigbee protocol used by Philips Hue light bulbs to communicate with each other. They say they were able to retrieve the encryption keys used to authenticate messages exchanged between the bulbs.

Many other attacks have taken place, with consequences of varying significance. Table 4.4 presents the most well-known attacks.

Year	Attack	Description	Consequences
2018	Alert (TA18-074A)	IS-CERT alert about an attack on US infrastructure, multiple attack path, including social engineering techniques	Information retrieval on ICS
2017	BrickerBot	Similar to Mirai with permanent destruction of contaminated objects	Attacks and deactivation of objects

Year	Name	Description	Impact
2017	TRITON	Attack on SIS security automatons (Triconex), Remote Access Trojan (RAT)	Shutdown of the installation, potential industrial disaster
2017	WannaCry	Massive attack affecting more than 300,000 workstations using a Windows flaw and performing data encryption, then a ransom request, virus spreading via a Windows flaw (EternalBlue)	Financial losses (ransom), production shutdown
2017	Petya	Attack on accounting software in Ukraine. Same vulnerability as Wannacry	Financial losses (ransom), production shutdown
2016	Philips Hue	Attack demonstration with takeover of a Philips Hue lamp network Vulnerability protocol and physical IoT attack	Demonstration of real scale vulnerability
2016	Mirai or DYN	Attack of Dyn servers by DDoS contamination of connected objects (camera, DVR players, etc.) using identifiers by default launch of a DDoS (request flooding) attack on servers translating site names into IP addresses	Internet blocking, name servers no longer providing services
2016	Lappeenranta Building attack	Attack on the heating system of a building in Finland (managed by Valtia), DDoS	Loss of heating
2015	BlackEnergy	Power outage for 6 h affecting 230,000 people in Ukraine, Spear phishing email to install a Trojan	Power supply failure
2015	German Steel Mill Cyber attack	Takeover of the control system of a blast furnace that generated massive damage, Spear Phishing email and Trojan	Physical damage

2014	DragonFly	Attacks against energy companies by compromising ICS equipment, Remote Access Trojan (RAT): Havex/Energy bear Email (pdf), Watering hole attack	Sabotage
2014	Sandworm	Attack on General Electric and Siemens software Zero Day Vulnerability Windows CVE 2014 4114 (OLE exec)	Sabotage
2012	Telvent Canada attack	Access to the administration tools of the control system Malware distribution via social engineering	Information theft SCADA software
2011	Night Dragon	Extraction of confidential information relating to oil and gas projects. Social engineering and root control	Data theft
2011	Duqu	Parts of the code almost identical to Stuxnet, designed only for industrial espionage without containing destructive functions	Data theft
2010	Stuxnet	Several years of infiltration into the Natanz uranium enrichment complex, damage to more than 900 uranium enrichment centrifuges Complex attack exploiting Windows and Step7 vulnerabilities	Degradation of centrifuges
1999	Gazprom (Russia)	Takeover of the distribution panel controlling gas flows from pipelines Trojan and internal complicity	Stopping of production
1982	Pipeline attack	Attack on a pipeline in Siberia, Trojan on SCADA	Gas pipeline explosion

Table 4.4. *Attacks on ICS*

4.11. Some statistics

Table 4.5 presents the most frequent attacks identified by BSI 54 (Federal Office for Information Security 2016). It should be noted that those using the human factor are in the lead. The incidents reported by sector in the United States by CERT in 2014 are shown in Figure 4.9. Of these, 55% correspond to APT attacks.

These incidents, although limited in number, are increasing sharply according to Kaspersky (Kaspersky Lab 2018).

No.	Top 10 2016	Top 10 2014
1	Social engineering and phishing	Malware via Internet Infection and Intranet
2	Malware spread via removable media and external hardware	Malware spread via removal media and external hardware
3	Malware via Internet Infection and Intranet	Social engineering
4	Intrusion via remote access	Human error and sabotage
5	Human error and sabotage	Intrusion via remote access
6	Control components connected to the Internet	Control components connected to the Internet
7	Technical failures and cases of force majeure	Technical failures and cases of force majeure
8	Compromise of Extranet and Cloud components	Compromising smartphones in the production environment
9	(D)DoS attacks	Compromise of Extranet components and Cloud components
10	Compromise of Extranet and Cloud components	(D)DoS attacks

Table 4.5. *The most frequent attacks*

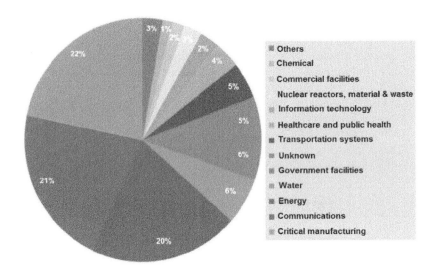

Figure 4.9. *Sectors affected by attacks. For a color version of this figure, see www.iste.co.uk/flaus/cybersecurity.zip*

5

Vulnerabilities of ICS

5.1. Introduction

The success of an attack on a computer system depends on at least one vulnerability being exploited; it can be technical, human or organizational.

Very often, vulnerability analysis of an industrial control system (ICS) is limited to a technical aspect. While the number of vulnerabilities of this type is often significant, they do not explain most attacks on their own, as shown by analysis of feedback. Fixing these technical vulnerabilities is necessary, but not sufficient.

The analogy with the security of a building makes you realize this: to ensure security, it is important to have good quality locks and to have secured the windows; this is the technical aspect. The building use policy should define rules related to closure, such as closing times. It is then necessary for users to lock the door, this is the human aspect. They must therefore be made aware of the importance of the procedure and, if necessary, trained in use of the locking system. To complete the whole, it is then necessary to plan to check that the rule is applied and, possibly, to plan a systemic lock when shifts end by a member of security staff.

An analysis of the building security vulnerabilities will begin with an analysis of the technical characteristics (lock quality), which should be as comprehensive as possible. This will require first identifying the different potential entry points for a burglar. In a second step, the organizational aspect will be checked: has a coherent policy been defined? Otherwise, an organizational vulnerability exists. Then, users will have to be informed or trained. Finally, it will be a question of verifying that the prescribed policy is indeed the one that is being implemented.

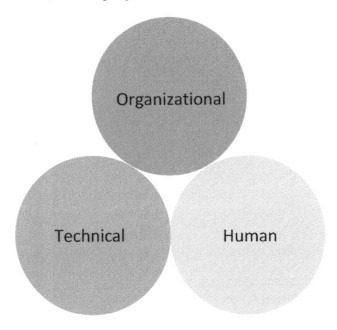

Figure 5.1. *The different types of vulnerabilities to consider*

This chapter then presents a generic approach to analyze the vulnerabilities of an ICS. These systems offer a number of potential entry points, which define the attack surface, presented in the next section, and which are useful to know in order to implement the appropriate measures.

Finally, the vulnerabilities of conventional industrial systems and Industrial Internet of Things (IIoT) systems are studied, and methodological or technical tools to analyze the vulnerability of an ICS system are presented.

5.2. Generic approach to vulnerability research

A review of the ICS architecture and feedback shows that an attack most often takes place as follows:

1) the first step consists of entering the industrial network (the one to which the ICS equipment is connected:

    i) from outside the industrial network (IT network, Internet, remote access):

        - after introducing a Trojan horse or a backdoor through social engineering;

        - through network protection vulnerability such as, for example, lack of network separation, vulnerability of network equipment, etc.;

        - via an external workstation used for remote maintenance, and on which a Trojan horse has been installed;

    ii) via a back door installed during manufacture or integration, with or without the knowledge of the intervener or manufacturer;

    iii) via removable media or contaminated mobile equipment (laptop, tablet, mobile phone);

    iv) from unsecured local wireless access;

    v) via a wired network connection used by an internal attacker.

At the end of this step, the attacker has equipment connected to the industrial network and able to discover other equipment, spy or corrupt exchanges.

2) the second step is discovery of the network, using scanning software from one of the workstations with access to the network;

3) the third, optional step consists of installing a Trojan horse or back door on other equipment;

4) the last step is to perform malicious actions on Supervisory Control And Data Acquisition programmable logic controller (SCADA PLCs), servers and supervision stations (espionage, data corruption, stopping or slowing down, direct dangerous actions on the physical process), directly or through modification of programs, more particularly:

    i) on servers and supervision stations, in displaying an incorrect view (actual values are not displayed and/or actual commands are not sent to PLCs);

    ii) on PLCs, by modifying data or programs to create system malfunctions or damage or to disconnect safety functions;

    iii) on historian databases.

A generic attack tree can summarize this approach (Figure 5.2).

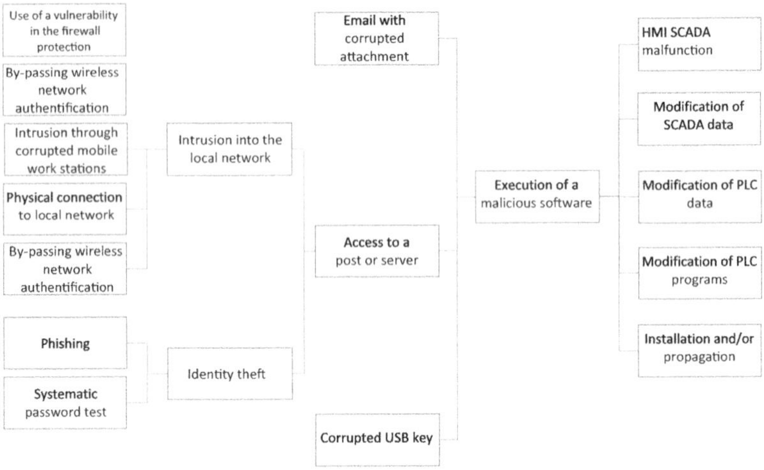

Figure 5.2. *Generic diagram of an attack*

To secure an ICS, it is necessary to take appropriate measures to prevent each of these steps from being carried out.

5.3. Attack surface

The attack surface (Manadhata and Wing 2011) of a system is defined as:

– a description of all the points through which an attacker can enter;

– the list of security measures;

– the list of targets to be targeted.

The generic attack surface of an ICS is shown in Figure 5.3. It shows intrusion points, targets and measurements. It should be noted that most of the targets are used as relays that allow new actions to be implemented in order to reach the final target of automata directly controlling the physical process.

In summary, the entry points of an ICS are as follows:

– poorly configured firewalls between the Information Technology (IT) and Operation Technology (OT) network;

Vulnerabilities of ICS 125

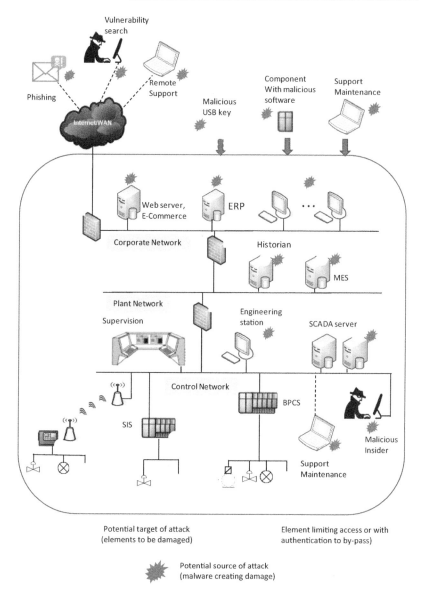

Figure 5.3. *Attack surface of an ICS. For a color version of this figure, see www.iste.co.uk/flaus/cybersecurity.zip*

– demilitarized zone (DMZ) servers (see Chapter 10), which can be accessed from the external network or IT to introduce malware and can be used as relays;

– all workstations outside the OT network that are connected to it via a virtual private network (VPN) (or without any particular protection, which is worse): if these workstations are corrupt, they are entry points;

– USB keys and other removable memory devices;

– mobile computers or workstations connected by Wi-Fi that can be corrupted, when they are outside the OT network;

– workstations in the OT network that can access the Internet or an external network, and can be corrupted by malware;

– the workstations in the OT network receiving emails;

– PLCs and other devices that connect to insufficiently secure update sites;

– workstations, equipment and software from untrustworthy suppliers;

– equipment with security breaches due to faulty design or implementation;

– unsecured physical access points.

As it can be seen, there must be a "physical" contact or logical link to exploit a vulnerability. The search for vulnerabilities in an installation and their correction are fundamental steps in risk control. The elements that can be affected, and that must be analyzed, are those presented in Figure 5.3, which shows the targets and elements that can be used as relays to conduct attacks.

5.4. Vulnerabilities of SCADA industrial systems

Before presenting the most exhaustive list of ICS's vulnerabilities, this section presents the vulnerabilities that are most often identified during the analyses.

From a global perspective, the main technical vulnerabilities used are related to:

– an unsecured architecture, allowing access to vulnerable equipment, or to modify traffic based on vulnerable protocols;

– poorly configured firewalls;

– the use of unreliable protocols that are easy to spy on and corrupt;

– ICS equipment without authentication capabilities;

– software vulnerabilities that are not patched with regular updates;

– weak authentication (default password, shared, weak, etc.).

At supervision level (level 2 of Purdue's model), vulnerabilities are fairly classic ones related to servers and workstations. The specific point of ICS architectures is the use of unsecured protocols, such as the Modbus protocol.

For level 1, where the PLCs are located, the vulnerabilities concern, on the one hand, aspects related to the equipment itself and, on the other hand, the mechanism that allows users to communicate with the PLC and to use it. From a technical point of view, a PLC has a central unit and a firmware or operating system that ensures basic operation. This operating system has vulnerabilities. A communication is made with the SCADA server and the engineering stations in order to transfer:

– the user program (in one of the IEC 61131-3 languages, see Chapter 1);

– the configuration;

– commands to control the operating mode (stop, run, reset, etc.);

– control commands sent from the SCADA workstation;

– the sending of data from the PLC to the SCADA and historian servers.

The security of this communication is based on:

– a user authentication mechanism, integrated into the equipment;

– a rights management policy;

– the management of user IDs.

Vulnerabilities may occur for each of these functions.

Finally, at level 0, if the Transmission Control Protocol (TCP)/Internet Protocol (IP) protocol is used with devices running with an embedded firmware or an embedded operating systems, there are vulnerabilities related to these two elements.

In IIoT architectures, the protocols used are secure: vulnerabilities result from errors in implementations or key management flaws. Note that devices are easier to access physically, so some attacks based on reverse engineering or electronic means (*Side Channel Power Analysis*, for example) can break some protection mechanisms.

The graph presented in Figure 5.4, obtained from a Kaspersky Labs study in 2017, provides an idea of the distribution of vulnerabilities: network and protocol aspects are important.

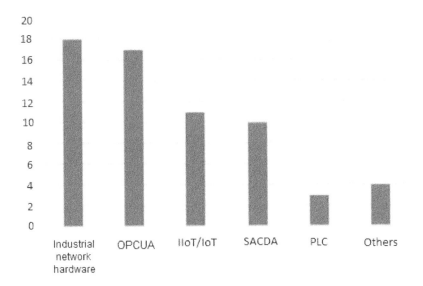

Figure 5.4. *Distribution of identified vulnerabilities*

5.5. Vulnerabilities of IoT industrial systems

In general, in addition to the vulnerabilities listed above (except for insecure protocols), vulnerabilities specific to IIoT equipment are due to their characteristics:

– the dispersion of elements leads to physical vulnerability;

– the perimeter of the elements constituting the system is poorly known;

– for systems with a low computing power, it is difficult to implement cryptographic functions, so exchanges are vulnerable;

– as systems are intended to evolve by adding or removing equipment, this operation must be carried out in a simple and flexible way, while remaining secure. In this phase, a vulnerability during key exchange may exist in the absence of asymmetric encryption with certificates, or if the root private key is known;

– the devices may be difficult to update securely;

– the similarity of systems can allow large-scale attacks when a flaw is found.

More concretely, the specific vulnerabilities are in hardware, system software and communications during the addition of an IIoT to a network.

At the hardware level, given the limited computing capacities and available energy, which may also be limited, vulnerabilities are due to the following:

– electronic analysis capabilities, using variations in power consumption (Power Side Channel Attack) (Mangard et al. 2007) to obtain the encryption keys;

– a possible lack of energy following incessant solicitations (Raymond et al. 2009);

– an injection, at the time of manufacture, of components to disrupt normal behavior, such as, for example, maintaining a processor's privilege bit at a fixed value (Yang et al. 2016).

At the system software level (firmware, OS and possibly privileged level applications such as a virtual machine), vulnerabilities are due to the following:

– less sophistication: for example, there is no memory management unit on simple microcontrollers, which is a problem for process isolation;

– access control management may be poorly designed and have inadequate granularity (Fernandes et al. 2016);

– software update is difficult: it can require a shutdown of the system under control, a reset and impose a long time for transfer, which is a problem for IoT equipment with few energy resources, especially those with energy harvesting.

IIoT systems often include equipment for local information processing (fog computing). This equipment can take various forms such as gateways with advanced processing capabilities or a central platform in the case of a vehicle.

This equipment presents vulnerabilities that should not be neglected in a risk analysis.

5.6. Systematic analysis of vulnerabilities

Analysis of system vulnerabilities is an important step in the security process. Two types of approach exist:

– the first is based on an audit-type analysis, using a checklist, which makes it possible to go beyond technical vulnerabilities;

– the second is based on software that automatically detects the various devices' vulnerabilities.

The NIST guide (Stouffer et al. 2015) or the ANSSI guides (ANSSI 2013b) provide examples of generic vulnerabilities to perform a checklist. Figure 5.2 shows the most common ones. This checklist makes it possible to carry out an initial identification of risks on an installation.

Figure 5.5. Steps of a CSET analysis

One of the most commonly used scanning tools to perform a vulnerability scan for an ICS is the Cyber Security Evaluation Tool (CSET), created by the ICS-CERT team (where CERT is Computer Emergency Response Team). It provides a consistent step-by-step approach (Figure 5.5) to evaluate security measures and practices, comparing them to the requirements identified in one or more industry-recognized standards (Table 5.1).

The first step consists of choosing a reference system from among those proposed (Table 5.1). Second, the security assurance level (SAL) is determined by the answers to questions about the potential consequences of a successful cyber-attack. It can be selected or calculated and provides a recommended level of security necessary to protect the system from a feared event.

It is then possible to draw a diagram of the installation. This can be imported from the GRASSMARLIN software (Chapter 10).

CSET then generates questions using the network topology, the security standard chosen and the level of SAL chosen. The analyst can select the best answer to each question using the organization's actual network configuration and the security policies and procedures implemented (Figure 5.6). Each question can be combined with background information provided by the analyst to support the answer.

The last step is the analysis of the results. CSET provides a dashboard and various possibilities for filtering results.

CFATS Risk Based Performance Standard (RBPS) 8: Chemical Facilities Anti-Terrorism Standard, Risk Based Performance Standards Guidance 8 – Cyber, 6 CFR Part 27
DHS Catalog of Control Systems Security: Recommendations for Standards Developers, Revisions 6 and 7
DoD Instruction 8500.2 Information Assurance Implementation, February 2, 2003
ISO/IEC 15408 revision 3.1: Common Criteria for Information Technology Security Evaluation, Revision 3.1
NERC Reliability Standards CIP-002-009 Revisions 2 and 3
NIST Special Publication 800-82 Guide to Industrial Control Systems Security, June 2011
NIST Special Publication 800-53, Recommended Security Controls for Federal Information Systems Rev 3 and with Appendix I, ICS Controls
NRC Regulatory Guide 5.71 Cyber Security Programs for Nuclear Facilities, January 2010

Table 5.1. *Standards supported by CSET*

Figure 5.6. *Answer to CSET questions screen. For a color version of this figure, see www.iste.co.uk/flaus/cybersecurity.zip*

Physical vulnerabilities
– No physical access control to the facility
Possible access for people who can destroy equipment or introduce malicious systems
– Unsecured ports (USB and Ethernet):   - USB port   - Ethernet cable   - Keyboard connection   - JTAG (test)
– No control of environmental factors
Temperature, humidity, rodents, floods, fire, etc.
– No emergency power supply
– No protection system against electromagnetic interference, overvoltages, static electricity

Organizational vulnerabilities
– No security policy for ICS or inappropriate policy
The vulnerabilities of an ICS are often due to the fact that there is no specific security policy for these systems, or that it does not manage all aspects, such as mobile computers.
– No definition of responsibilities for security issues between supplier, integrator and operator
Very often, the information security policy does not clearly define, as far as industrial systems are concerned, the responsibilities devolved to the various stakeholders, namely the supplier, the integrator and the operator.
– No risk management system in place
No risk management approach has been implemented.
– No risk assessment
The ICS has not been the subject of a cybersecurity risk analysis.
– Unmanaged external stakeholders
May be a source of threat, voluntarily or not.
– No consideration of cybersecurity in projects
Cybersecurity aspects are not taken into account when building new installations or extensions.
– No management of suppliers and service providers
Suppliers and service providers must be audited or certified. In most cases, cybersecurity is not taken into account with regard to external providers, including those developing PLC software or SCADA configurations. Remote access may also have been opened.
– Lack of procedures or guides for the implementation of ICS equipment
– Lack of configuration management policy
The organization does not have an inventory of hardware and software with versions and update status.
– Lack of user access and authentication policy
Roles are not defined, access controls are not managed.
– Lack of review or audit of the effectiveness of security measures deployed for the ICS
Procedures and planning must exist to ensure that the measures are implemented correctly.

User access
– Inadequate password management (default password, rarely changed or weak)
– Inadequate user account management (shared passwords, no staff turnover management, administrator mode operation)
Human factor
– Lack of awareness of the risks of cyber attacks
The risk of cyber-attacks is perceived as remote and, more often than not, the availability of the installation and ease of use (remote access) take precedence.
– Lack of awareness of the risks associated with social engineering
Most attacks against industrial sites have been carried out using this vector, yet many users do not take adequate precautions.
Architecture and network

– No inventory or mapping:
   - No list of computer workstations
   - No list of equipment
   - No controls on connected laptops
   - No list of network equipment and topology
   - No list of allowed logical flows between applications and units
   - No list of external connections (IT network or Internet)
The properties to be protected have not been identified, the perimeter is not known.

– Unadequate partitioning design:
   - Poorly defined or undefined perimeter
   - Incorrect design of the zones
   - No partitioning between the industrial system and the management information system
   - No partitioning between the different parts of the industrial system
   - No partitioning of the IT administration of the system

– Firewall or network equipment not properly configured or updated
The rules are poorly designed, poorly updated or the firmware itself is not up to date.

– Use of vulnerable protocols

ICS protocols are very vulnerable, they should only be used in the OT area and secured with adequate firewall rules.

– Access for remote maintenance not secured

Often used for updates by service providers and for remote isolated equipment (e.g. measuring station), but often opens security breaches.

– Unmanaged mobile terminals

Nowadays, including in ICS, more and more features are available from the phone or tablets, often with personal devices (Bring Your Own Device, BYOD). Security is not always controlled for these positions.

– Use of VPN without caution

Most users think that using a VPN provides absolute security, although this is not the case. It's just a way to transport information securely. But the source may be corrupted and the destination remains vulnerable.

– File sharing with few levels of control

Files are often shared without controls or access rights, which facilitates attacks.

– Insufficiently secure Wi-Fi networks

Protocols are not always well selected and are often visible from outside the site. An employee who no longer belongs to the company can, for example, connect to it.

Equipment: Office automation station (SCADA stations, touch screens), mobile terminals, sensor and actuator configuration tools, PLC programming station, remote terminals, network equipment (switches, routers, wires), PLC, IIoT devices, sensors, actuators

– Use of vulnerable equipment and workstations:

- Configurations are not hardened (open software ports, unnecessary modules, etc.).

- Software not tested for security

- No antivirus or not up to date

- Development environment not hardened

- Development tools on programming stations not uninstalled

Off-the-shelf commercial components (COTS) are increasingly used in projects to reduce design, manufacturing and maintenance costs. This may be software integrated into a particular project (e.g. Office package, web server, video player, etc.), or hardware components. These components facilitate the implementation of generic attacks.

– Inadequate vulnerability management:
  - No software vulnerability management device (embedded applications and software)
  - Security patches not regularly applied
  - No vulnerability monitoring

– Lack of control of the PLC configuration:
  - No integrity check of the configuration and/or programs
  - No authentication when uploading PLC program
  - No configuration backup

Monitoring and response
– No supervision of cybersecurity events *Cybersecurity incidents are currently very rarely detected in order to generate an alert on the supervision screens. Logging of these events is also very limited. An attacker can therefore act with little risk of being spotted.*
– Lack of business continuity plan *Current business continuity plans very rarely take into account events related to a malicious attack on the industrial information system. Operators are often helpless in the event of a malicious attack.*
– Lack of a crisis management training program *In the event of an incident, staff are not trained to react.*

Table 5.2. *Checklist for examining vulnerabilities*

5.7. Practical tools to analyze technical vulnerability

Vulnerability analysis based on a checklist can be complemented with an analysis of the existing installation by software tools whose objective is to find effective vulnerabilities.

Two approaches exist:

– the first consists of consulting a database of identified vulnerabilities, using the hardware and software inventory (type and version number);

– the second is to use penetration test tools to identify vulnerabilities in the installation. This second approach should of course be used with caution with an operational installation, given the disruption it may produce.

5.7.1. Databases and information sources

ISO 27005 or NIST SP 800-53 (NIST 2014) provides standard lists of generic vulnerabilities.

CERTs provide a list of actual vulnerabilities encountered on different hardware and software, and they issue alerts. The information is accessible via a website. There are CERTs in different countries, as well as a site specific to ICS. It is managed by the National Cybersecurity and Communications Integration Center (NCCIC), a division of the Department of Homeland Security (DHS). It identifies incidents affecting ICSs, analyzes their origins and proposes solutions. Other databases exist. Here is a list of the main sites:

– National Cybersecurity and Communications Integration Center (NCCIC) *Industrial Control Systems* (ICS-CERT)[1];

– United States Computer Emergency Readiness Team (US-CERT)[2];

– Government Centre for Monitoring, Alerting and Response to Computer Attacks, CERT_FR[3];

– National Vulnerability Database[4];

– Common Vulnerabilities and Exposures Database[5];

– Security Focus[6];

– Exploit Database[7].

5.7.2. Pentest tools

Penetration testing is a common practice of security professionals to find system vulnerabilities before attackers do. They provide a real view of actual technical vulnerabilities, and are therefore a very useful complement to audit-based vulnerability analyses, even though they focus on a subset of the vulnerabilities in an installation.

---

1 Available at: https://ics-cert.us-cert.gov/.
2 Available at: https://www.us-cert.gov/.
3 Available at: https://www.certa.ssi.gouv.fr/.
4 Available at: https://nvd.nist.gov/.
5 Available at: https://cve.mitre.org/.
6 Available at: https://www.securityfocus.com/.
7 Available at: https://www.exploit-db.com/.

Penetration tests can be implemented as part of simulation exercises by dividing the members of the security team into two groups: a red team and a blue team. The red team plays the role of a hostile force, and the blue team defends the attacked system. The objective of the red team is to identify and exploit security weaknesses, both technical and organizational, while the blue team aims to defend the system by finding and fixing vulnerabilities, and responding to successful violations.

A number of commercial and open-source tools exist to perform penetration tests. Let us quote, for example:

– OpenVAS[8], Nessus' open source fork, is a set of tools that provides a complete and powerful vulnerability management solution;

– Kali Linux[9] is a Linux distribution dedicated to penetration testing. It includes many tools including OpenVAS, nmap, Metasploit;

– Linux-soft-exploit-suggester and Windows-Exploit-Suggester[10] are more specific tools for analyzing a workstation's software vulnerability.

Figure 5.7. *Example of OpenVAS analysis. For a color version of this figure, see www.iste.co.uk/flaus/cybersecurity.zip*

---

8 Available at: http://www.openvas.org/.
9 Available at: https://tools.kali.org/.
10 Available at: https://github.com/GDSSecurity/Windows-Exploit-Suggester.

5.7.3. Search engines

There are also search engines that scan the Internet for vulnerable connected objects.

Shodan[11] is the most popular specialized search engine in the search for electronic devices connected to the Internet. Shodan searches the entire Internet address space and captures information about each device it finds.

This information may include IP addresses, ports, banner messages, location, webcam images, etc. Shodan offers the possibility to search its database for various devices: servers, desktops and laptops, PLCs, HMIs, etc. It is possible to use filters, which makes it possible to carry out very specific searches.

Port: Filters based on port provided
Country: Searches for a specific country, using a two-letter code (e.g. US)
Net: Searches a network block using CIDR notation
Hostname: Searches for matching hostnames
OS: Searches based on the provided operating system
City: Searches a specific city
Geo: Searches specific coordinates
Before/after: Searches within a timeframe
Product: Finds specific product(s)
Version: Displays certain version(s)

Table 5.3. *Shodan filters*

Censys[12] is not as well known as Shodan as an IoT search engine, but it is also interesting. The features of Censys and Shodan are very similar, but Censys has features that distinguish it. For example, it provides more detailed information on certain protocols.

Shocens[13] allows both engines to be interrogated simultaneously.

---

11 Available at: www.shodan.io.
12 Available at: https://censys.io/.
13 Available at: https://github.com/thesubtlety/shocens.

6

Standards, Guides and Regulatory Aspects

6.1. Introduction

In recent years, many standards and guides have been proposed in the field of information system security. Some of these standards propose an approach for risk management in line with ISO 31000: these are ISO 27000 standards. Other standards focus on industrial control system (ICS), such as IEC 62443 or the NIST SP 800-82 guide. Others have been developed for a particular field, such as electricity distribution or production, or the nuclear sector. This chapter presents the main standards (Figure 6.1), including the ISO 27000 family, National Institute of Standards and Technology (NIST) guides, the ANSSI approach, NERC CIP sector standards and International Atomic Energy Agency (IAEA) standards. The IEC 62443 standard is presented in Chapter 7, which focuses on it, while the standards for operational safety, the IEC 61508 family, are introduced in Chapter 8.

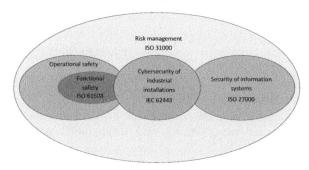

Figure 6.1. *Relations between the main standards*

6.2. ISO 27000 family

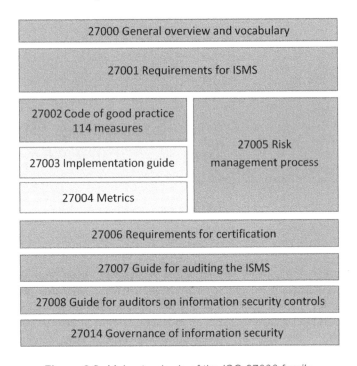

Figure 6.2. *Main standards of the ISO 27000 family*

The ISO 27000 family of standards defines good practices for information system security management. These have evolved over the years and are part of the general ISO 31000 framework, which describes the principles and guidelines for risk management, as well as the implementation processes at strategic and operational level.

The most important parts of this family include the following:

– ISO 27000, which provides an overview and defines the vocabulary;

– ISO 27001, which provides an approach for an organization to implement and improve the information security management system (ISMS) (see section 3.6 for more details), and normative requirements for the development and use of an ISMS;

– ISO 27002, which is a set of good practices for information security management. It proposes 114 security controls organized into 14 themes. It is designed for organizations wishing to select the necessary security measures as part of the process of implementing an ISMS such as the one described in standard 27001;

– ISO 27003, which is an implementation guide for ISMS;

– ISO 27004, which is a guide for the development of metrics for the implementation of ISMS;

– ISO 27005, which describes the information security risk management process (Figure 6.3) in accordance with ISO 31000. It is based on the general concepts specified in standards 27000 and 27001;

– ISO 27006, which defines the requirements for the accreditation of organizations for the certification of ISMS;

– ISO 27007, which is a guide for the audit of an ISMS;

– ISO 27019, entitled "Information technology–Security techniques–Information security management guidelines based on ISO/IEC 27002 for process control systems specific to the energy utility industry", which addresses cybersecurity of energy distribution systems. It is intended to help interpret and apply the ISO/IEC 27002 standard for this type of industry. The standard was first published as a technical report in 2013. It was revised in October 2017 to become a complete international standard harmonized with the 2013 version of ISO/IEC 27001 and 27002, in conjunction with IEC standards TC 57 and TC 65 (IEC 62443-2-1) and IEC SC45A (IEC 62645);

– ISO 27031, which describes the concepts and principles for preparing information and communication technology for business continuity;

– ISO 27032, which concerns the management of cybersecurity, in the sense of information security in a cyberspace (Internet network) context;

– ISO 27035, which concerns the management of security incidents. It defines a five-step process: preparation, incident identification, incident evaluation, response and feedback management;

– ISO 27799, entitled "Health informatics–Health information security management using ISO/IEC 27002", which was first published in 2008 and revised in 2016. It specifies guidelines for interpreting and implementing the 27002 standard in the field of health informatics and is complementary to it.

Among the important elements highlighted in these different standards, the concept of ISMS is described in section 3.6, the risk management process in section 3.3 and the security measures in standard 27002 are presented in Chapter 9 (Figure 9.3). In addition, the IEC 62443 standard, presented in Chapter 7, is aligned with ISO 27000 standards.

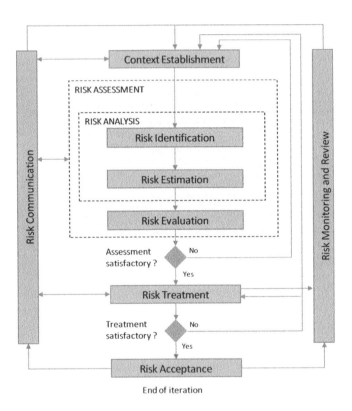

Figure 6.3. *Risk management process (ISO27005)*

6.3. NIST framework and guides

6.3.1. *NIST Cyber Security Framework*

The security of an installation is divided into several stages, ranging from the installation of protection to the development of reaction capabilities. These steps have been formalized by the NIST in a framework called the Cyber Security Framework (CSF) (NIST 2018). These steps are as follows:

– identification;

– protection;

– detection, during operation;

– the response to attacks;

– recovery.

For each phase, a number of areas have been identified (Figure 6.4). They include a list of measures to be implemented.

A version of this framework has been adapted for industrial production systems (Stouffer *et al.* 2017).

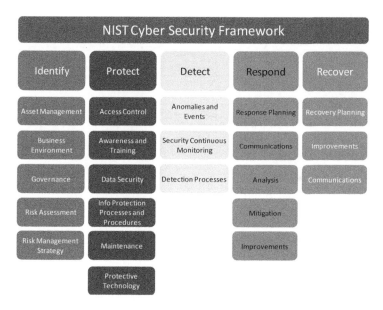

Figure 6.4. *Structure of the NIST framework. For a color version of this figure, see www.iste.co.uk/flaus/cybersecurity.zip*

6.3.2. *The guides*

In addition, the NIST publishes many guides, which include the following:

– the NIST SP 800-82 guide, which describes a comprehensive approach to securing ICS;

– the very comprehensive NIST SP 800-53 guide, which provides a list of security measures organized according to CSF headings;

– the NIST guide SP 800-30, which deals with risk assessment for IT systems;

– many ad hoc guides, such as NIST SP800-41, *Guidelines on Firewalls and Firewall Policy*.

IDENTIFY	ID.AM	Asset Management
	ID.BE	Business Environment
	ID.GV	Governance
	ID.RA	Risk Assessment
	ID.RM	Risk Management Strategy
	ID.SC	Supply Chain Risk Management
PROTECT (PR)	PR.AC	Identity Management, Authentication and Access Control
	PR.AT	Awareness and Training
	PR.DS	Data Security
	PR.IP	Information Protection Processes and Procedures
	PR.MA	Maintenance
	PR.PT	Protective Technology
DETECT (DE)	DE.AE	Anomalies and Events
	DE.CM	Security Continuous Monitoring
	DE.DP	Detection Processes
RESPOND (RS)	RS.RP	Response Planning
	RS.CO	Communications
	RS.AN	Analysis
	RS.MI	Mitigation
	RS.IM	Improvements
RECOVER (RC)	RC.RP	Recovery Planning
	RC.IM	Improvements
	RC.CO	Communications

Figure 6.5. *Structure of the NIST Framework. For a color version of this figure, see www.iste.co.uk/flaus/cybersecurity.zip*

The SP800-82 guide is dedicated to ICS. The approach it describes is part of a four-phase risk management process: scoping, assessment, response and monitoring, which corresponds to the process proposed in the NIST SP800-39 guide.

The scoping step sets the context and defines an approach with which the organization manages risks. This step produces a risk management strategy that translates into the risk assessment, response and monitoring steps.

Risk assessment is classic: it consists of identifying threats, impacts and estimating the level of each risk.

Risk response defines how the organization chooses to address the risks it faces: acceptance, mitigation, avoidance or transfer.

Risk monitoring involves periodic or continuous action to validate known sources of risk, identify new sources (external threats or internal environmental changes) and verify the implementation or validation of the effectiveness of actions chosen as part of the risk response.

The process is therefore similar to that described in Chapter 3.

The SP800-82 guide describes the steps of a management system for the security process of an ICS:

– development of an economic argument highlighting the stakes: potential benefits of a security management system, costs of potential damage, high-level view on the process to be implemented to manage security and the cost of the resources needed to implement this program;

– presentation of this argument to decision makers;

– setting up a multi-skilled team (IT staff, control engineer, operator, etc.);

– definition of the charter and its scope;

– definition of the policy and procedures specific to the security of ICS;

– inventory of ICS equipment and definition of security requirements in terms of CIA (confidentiality, integrity, availability);

– selection of measures to be implemented according to security requirements;

– risk assessment, to determine the desired level of protection (low, medium, high) and to determine countermeasures;

– implementation of measures.

The proposed measures are in line with the NIST SP 800-53 guide and include 18 themes (Appendix 3):

– access control;

– awareness and training;

– audit and accountability;

– security assessment and authorization;

– configuration management;

– contingency planning;

– identification and authentication;

– incident response;

– maintenance;

– media protection;

– physical and environmental protection;

– planning;

– personnel security;

– risk assessment;

– system and services acquisition;

– system and communications protection;

– system and information integrity;

– organization-wide security management;

– privacy controls.

Each theme includes several measures and submeasures, that are selected - or not, according to the desired level of protection (low, medium, high).

The proposed approach is consistent with the approach presented in Chapter 11. Only the first step (economic argument) is developed in more detail in the guide.

6.4. Distribution and production of electrical energy

6.4.1. NERC CIP

The NERC CIP (North American Electric Reliability Corporation Critical Infrastructure Protection) standard includes nine components and 45 requirements covering the security of power generation and transmission systems. It includes the protection of critical IT assets, as well as personnel and training, security management and disaster recovery planning.

PAC-001	Sabotage reporting
CIP-002	Critical cyber asset identification
CIP-003	Security management controls
CIP-004	Personnel and training
CIP-005	Electronic security perimeters
CIP-006	Physical security of critical cyber assets
CIP-007	Systems security management
CIP-008	Incident reporting and response planning
CIP-009	Recovery plans for critical cyber assets

Table 6.1. *Structure of the CIP standard*

The main objective of CIP-002 to CIP-009 (CIP-001 is not related to cybersecurity) is to protect the electricity distribution system from unwanted and destructive effects caused by cyber-terrorism and other cyber-attacks, including internal attacks.

Under the NERC CIP standard, organizations are required to identify critical assets and regularly conduct a risk analysis of these assets. Strategies for monitoring and modifying the configuration of critical assets must be defined, as must the rules governing access to these resources. In addition, NERC CIP requires the implementation of protection systems such as the use of firewalls to block vulnerable ports and tools to monitor cyber-attacks.

Security event monitoring systems must be deployed, and organizations must have comprehensive contingency plans for cyber-attacks, natural disasters and other unforeseen events.

The CIP standard is one of the 14 mandatory standards of the Federal Energy Regulatory Commission (FERC) in the United States. Penalties for non-compliance with NERC CIP may include fines, sanctions or other actions against the entities concerned.

FERC's objective with this standard is to ensure that the North American power grid will not fail due to cybercrime.

6.4.2. IEC 62351

The IEC 62351 standard is a standard developed by WG15 of the IEC TC 57 group. It has been designed to manage the security of the TC 57 group protocol series, including the 60870 series, used by electrical power distributors (Chapter 2). TC 57 is responsible for developing standards for the exchange of information on energy supply and related systems, including energy management systems, SCADA control systems, distribution automation and remote protection. The various security objectives of IEC 62351 include authentication of data transfer via digital signatures, ensuring authenticated access, prevention of wiretapping, prevention of identity reading and misuse, and intrusion detection (Table 6.2).

IEC 62351-1-1	Introduction to the standard
IEC 62351-2	Glossary of terms
IEC 62351-3	Security for any profiles including TCP/IP
IEC 62351-4	Security for any profiles including MMS (e.g. ICCP-based IEC 60870-6, IEC 61850, etc.)
IEC 62351-5	Security for any profiles including IEC 60870-5 (e.g. DNP3 derivative)
IEC 62351-6	Security for IEC 61850 profiles
IEC 62351-7	Security through network and system management
IEC 62351-8	Role-based access control
IEC 62351-9	Key management
IEC 62351-10-10	Security architecture
IEC 62351-11-11	Security for XML files

Table 6.2. *Structure of standard 62351*

6.4.3. *IEEE 1686*

In the field of electrical distribution, there is also the IEEE 1686 standard for intelligent electronic devices (Chapter 1) cybersecurity (IEEE 2013).

6.5. Nuclear industry

6.5.1. *The IAEA technical guide*

A first guide, entitled Computer Security at Nuclear Facilities (International Atomic Energy Agency 2011), provides specific advice to key nuclear facilities on implementing a computer security program and evaluating existing programs.

A second guide, entitled Computer Security of Instrumentation and Control Systems at Nuclear Facilities (International Atomic Energy Agency 2015), describes security measures for instrumentation and control (I&C) systems.

These guides recommend that the operator define IT security requirements based on a gradual approach, based on the risk level, and taking into account the following elements:

– the importance of the I&C system's functions for safety and security;

– the threats identified and assessed for the installation;

– the attractiveness of the I&C system for potential opponents;

– the I&C system vulnerabilities;

– the operating environment;

– the potential consequences that can result directly or indirectly from a compromise of the system.

Such an approach can be based on the results of a risk assessment (International Atomic Energy Agency 2016).

The concept of zone is introduced in this standard, as well as the principle of defense in depth. A step-by-step approach is proposed, based on the diagram in Figure 6.6. For each level, specific measures are defined.

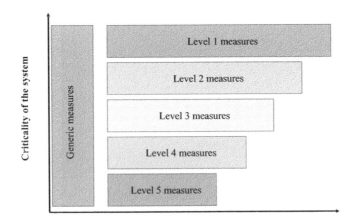

Figure 6.6. *Gradual approach. For a color version of this figure, see www.iste.co.uk/flaus/cybersecurity.zip*

These guides follow the same approach as that proposed in IEC 62645 described in the following section.

6.5.2. *IEC 62645*

This standard was developed by the IEC/SC45A group in charge of the instrumentation, control and power supply aspects of nuclear installations. This group works in a coordinated way with the IAEA. The IEC 62645 standard was published in 2014. It has been designed to be consistent with ISO 27001 and ISO 27002. Overall, standard 62645 is mainly structured into three parts:

– a first one dealing with the lifecycle of security at the level of the security program;

– a second one dealing with the lifecycle of security at system level;

– a third describing security measures by theme: security policy, asset management, human aspects, etc.

According to this new standard, IT system security must be based on a graduated approach, according to the following principles:

– three security levels (called security levels S1, S2 and S3) are defined in the standard. Security measures cannot be defined individually for each system, as this would lead to a large amount of studies (and cost) and to many problems for connecting communicating systems;

– the systems must be considered from a functional point of view and have a level of security based on their possible direct or indirect impact on the security and availability of the installations;

– the generic measures given must be adapted to each level in order to effectively protect the systems of each level considered.

Some generic characteristics in development of the graduated approach are as follows:

– security programs must be developed according to the level of system software development, with the establishment of a secure development and of a secure operating environment during the various phases of the software lifecycle;

– a similar level of security must be achieved for all systems with the same security requirement, regardless of their designer and developer;

– interfaces between systems with different security levels must be specifically addressed;

– interfaces must be secured, but must not prevent functional transmission.

6.6. Transportation

6.6.1. *Vehicles*

The SAE J306 standard, which concerns motor vehicles, was developed with the ISO 26262 (Chapter 7) operational safety standard in mind. It describes a structured process to reduce the probability of a successful attack.

The principles of the approach are as follows: consider the use of functionality by vehicle owners, implement cybersecurity in the design and engineering phases, implement cybersecurity in development and validation, implement cybersecurity in incident response, and consider cybersecurity when the vehicle owner changes.

6.6.2. *Aeronautics*

Nowadays, with increasing connectivity, security-based security is crucial. RTCA Safety Standard DO-326A (2014) applies to aircraft and aircraft systems. It only deals with security aspects that could have an impact on flight safety. This standard specifies a process for top-down risk assessment with a generic set of activities, and is compatible with other industry standards dedicated to the certification of aircraft systems.

This standard separates the IT security and safety aspects, with feedback only, from safety to security.

6.7. Other standards

6.7.1. *National Information Security Standards*

BS7799 (2002) is a British standard describing good practices for information security management, consisting of three parts. It provides detailed and structured coverage of security issues. It has been incorporated into standards 27001:2013 and 27002:2013. It was adopted by ISO as an ISO 17799 standard in 2000.

IT-Grundschutz is a German security standard. It is one of a series of guides published by the German Federal Office for Information Security (BSI), which describe "information security methods, processes, procedures, approaches and measures" based on ISO/IEC 27001:2013.

This standard provides a guideline for conducting a risk analysis and includes a large number of security controls to provide a relatively high level of protection, without having to perform a detailed risk analysis. The purpose of the IT-Grundschutz risk assessment method is to provide a qualitative assessment; it includes the identification, analysis and evaluation of security incidents that could be harmful to the company.

Other examples are the Swedish standard SS627799, replaced by ISO 27001, or the GB/T22080-2008 standard in China corresponding to ISO 27001.

6.7.2. *Operating safety standards*

Standard 61508 (and its derivatives) describes the functional safety approach to ensure that a system presents a risk below a set threshold for the various hazards it may encounter. It is discussed in Chapter 8.

6.8. ANSSI's approach

In a general manner, this approach follows the risk management process approach described in ISO 27005, adding certain aspects specific to ICS. It is detailed in ANSSI (2013a, 2013b) and can be summarized as follows:

– description of the installation, functional and/or physical, showing the physical components and the components of the industrial information system. Section 10.1 describes the approaches to achieve this description;

– mapping of industrial information system (IIS) components, physical and logical, and hosted applications;

– partial risk analysis and classification of the installation with the approach detailed below (Figure 6.7);

– identification of countermeasures to be implemented depending on classification.

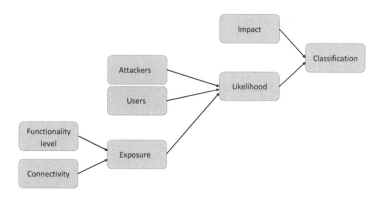

Figure 6.7. *General approach to classification*

This approach introduces the notion of class for a system, which allows the risk level of a facility to be taken into account. The classes are defined as follows:

– class 1: these are industrial systems for which the risk or impact of an attack is low. All measures recommended for this class must be applicable in complete autonomy. This level is the default level for any installation, and the proposed measures are similar to basic good practices;

– class 2: these are industrial systems for which the risk or impact of an attack is significant. There is no public control for this class of industrial system, but the responsible entity must be able to provide evidence that adequate measures have been put in place in the event of verification or of an incident;

– class 3: these are industrial systems for which the risk or impact of an attack is critical. In this class, the obligations are higher and the conformity of these industrial systems is verified by public authorities or an accredited body.

The level assessment is based on the rating of a number of installation characteristics (Figure 6.7). Combination rules are used to obtain the class of the installation. It is a combination of likelihood and impact level, therefore homogeneous to a risk level. A detailed guide proposes a series of measures to be implemented to reduce the risk. They are presented in Appendix 4.

The approach to evaluate the class is as follows:

– determine the level of functionality, noted F (Table 6.6) and the level of connectivity C (Table 6.7) to determine the level of exposure noted E with a matrix (Figure 6.8, left matrix);

– determine the level of attackers A (Table 6.4) and the type of stakeholders (Table 6.5), noted I;

– calculate the likelihood: $V = E + \left\lceil \frac{A+I-2}{2} \right\rceil$ by rounding up to the next upper integer;

– determine the severity level (Table 6.3);

– determine the class (Figure 6.8, matrix on the right).

The level of measures to be implemented depends on the class obtained (ANSSI 2013a).

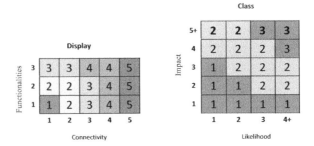

Figure 6.8. *Matrix to determine exposure level and class. For a color version of this figure, see www.iste.co.uk/flaus/cybersecurity.zip*

Level		Human Consequences	Environmental consequences	Consequences on the service
1	Insignificant	Reported accident without stopping or medical treatment	Limited and temporary violation of a rejection standard without legal reporting requirements to the authorities	Heavy impacts on 1,000 people
2	Minor	Reported accident with sick leave or medical treatment	Violating a discharge standard requiring reporting to authorities but without environmental consequences	Heavy impacts on 10,000 people. Disruption of the local economy
3	Moderate	Permanent disability	Moderate pollution limited to the site	Heavy impacts on 100,000 people. Temporary loss of major infrastructure
4	Major	One death. Permanent disability	Significant pollution or pollution external to the site Evacuation of people	Heavy impacts on more than 1,000,000 people Permanent loss of a major infrastructure
5	Catastrophic	Several deaths	Major pollution with lasting environmental consequences external to the site	Heavy impacts on 10,000,000 people Permanent loss of critical infrastructure

Table 6.3. *Severity level*

	Level	Description
1	Non-targeted	Viruses, robots, etc.
2	Hobbyist	People with very limited ressources, not necessarily a willingness to harm
3	Isolated attacker	Person or organization with limited resources but with some determination (e.g., a licensed employee)
4	Private organization	Organization with significant resources (e.g., terrorism, unfair competition)
5	State organization	Organization with unlimited resources and a very strong determination

Table 6.4. *Attacker level*

Level		Description
1	Authorized and controlled	All authorized participants are authorized and controlled. Unauthorized intervention is not possible
2	Authorized, and controlled	All authorized players are authorized, but at least some of the possible operations are not tracked. Unauthorized intervention is not possible
3	Authorized	There is no specific requirement for authorized intervenors but an unauthorized intervention is not possible
4	Not allowed	This category contains all industrial systems in which unauthorized intervention is possible

Table 6.5. *Types of stakeholders (noted I)*

Level		Description
F1	Minimum systems	This category includes industrial systems with only CIM 0 and level 1 elements[1] (control-command) excluding programming consoles, namely: – sensors/actuators; – remote inputs/outputs; – PLCs; – desks; – embedded systems; – analyzers.
F2	Complex systems	Complex systems. This category includes industrial systems containing only CIM level 0 to 2 elements (control and command and SCADA)
F3	Very complex systems	This category includes all industrial systems that do not fall into the first two categories. In particular, all systems: – with programming consoles; – with permanently connected engineering stations; – which are connected to a manufacturing execution system; – with centralized historian databases.

Table 6.6. *Level of functionality*

---

1 The CIM levels (Purdue model) are detailed in Chapter 1.

Level		Description
1	Isolated industrial system	Completely closed production networks
2	Industrial system connected to an IS	Production networks connected to the company's management information system, but without operations from outside the management information system being authorized
3	Industrial system using wireless technology	Industrial systems using wireless technology
4	Distributed industrial system with private infrastructure	A distributed system where the different sites communicate with each other through a private infrastructure (completely private or leased from a telecommunications operator), or with operations from outside or from a management network, such as remote diagnosis and maintenance
5	Distributed industrial system with public infrastructure	Similar to the previous category, except that the infrastructure used is public, such as that of a telecommunications operator. Example: Water distribution infrastructure

Table 6.7. *Connectivity level*

6.9. Good practices for securing industrial Internet of Things equipment

The Industrial Internet Consortium has proposed a set of good practices for securing industrial Internet of Things terminal equipment (Hanna *et al.* 2018), which are defined as components with computing capabilities and network connectivity. This can be, for example, a sensor or actuator in the world of car making, an embedded medical device in the world of healthcare or a pump or a flow sensor in the industrial world.

The proposed practices are based on NIST standards 800-53 and 800-82 (NIST 2014; Stouffer *et al.* 2015) and IEC 62443 (Chapter 7).

This guide defines three levels of security: basic, enhanced and critical. These levels correspond to security levels 2, 3 and 4 defined by IEC 62443 part 3-3. Security levels 0 and 1 of IEC 62443-3-3 are not covered, as they cover low security environments, which is not suitable for industrial environments connected to the Internet. The NIST SP 800-53r4 guide defines three levels of security in the same way.

The security levels are defined as follows:

– the basic security level (BSL) provides protection against "an intentional violation by simple means with limited resources", such as an ordinary virus;

– the enhanced security level (ESL) corresponds to a defense against "sophisticated means with moderate resources", such as the exploitation of known vulnerabilities in software or ICS;

– the critical security level (CSL) is adapted against attackers with "sophisticated means, extended resources", such as the ability to develop tailor-made zero day attacks.

Each piece of equipment must have an appropriate level of security, determined by risk analysis. Each security level has its own specific architecture (Figure 6.9). The following sections describe the elements of these architectures.

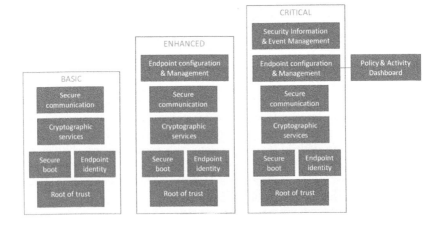

Figure 6.9. *Security profile*

6.9.1. *Trust base (root of trust)*

This first element corresponds to basic requirements to guarantee security. It is about providing functions to guarantee:

– the identity of the element (a unique identifier);

– the identity and integrity of the software and electronic components.

In practice, the object must have a secure element memory to store keys and certificates that are used to check the integrity of the software at start-up, to guarantee the identity of the object or to establish secure communications. This type of memory is similar to what is found in smart cards.

6.9.2. *Identity management (endpoint identity)*

Guaranteeing identity is a fundamental aspect. Support for a Public Key Infrastructure (PKI) is mandatory for all security levels. Protocols defined by an open standard for managing standard certificates (e.g. TSE) are used to automate the issuance, renewal, updating and revocation of certificates issued by an internal or external certification authority.

6.9.3. *Secure boot*

At start-up, the object performs a secure boot (or secure bootstrap):

– verification of the integrity of the firmware and software modules by comparing their fingerprints (with an SHA256 algorithm, for example) with those stored. Fingerprints can be encrypted with asymmetric encryption, the object having a public key and the provider of a public key;

– identification, using the certificates in memory, of the equipment with which it is associated.

6.9.4. *Cryptographic services*

The equipment must provide cryptographic services for data transport and storage. These services include:

– functions supporting the management of Public Key Cryptography Standards (PKCS) for asymmetric and symmetric encryption, hash functions and random number generators of adequate resistance (these numbers are the basis of encryption algorithms);

– implementations of validated cryptographic algorithms (NIST/FIPS standards);

– an ability to update these algorithms (in the event of progress in decryption algorithms, particularly quantum computing);

– interoperability of cryptographic key types and certificates between multivendor systems.

6.9.5. Secure communications

A stack of secure end-to-end communication protocols is required, including:

– support for scalable authentication protocols to authenticate the equipment;

– an encrypted communication medium equipment-cloud;

– an encrypted equipment to equipment communication support for key management, for example.

More details are given in Hanna *et al.* (2018).

6.9.6. Equipment configuration and management

This function is intended to allow a secure update of the firmware and operating system (OS). It must rely on PKCS standards for data encryption and validate the sources and destinations of updates via certificates.

This function is required for advanced and critical levels.

6.9.7. Activity dashboard and event management by a SIEM

Continuous monitoring of equipment requires:

– configuration control to detect unauthorized changes in firmware, OS and installed applications;

– application-level controls to detect and prevent unauthorized activities (e.g., the use of unsecured encryption, hashing algorithms) that compromise data confidentiality or integrity.

In addition, remote security policy management functions must be possible for Operation Technology (OT) operators, and selected security events must be communicated in an appropriate format to an SIEM (Chapter 10).

These functions are required for the critical level.

6.10. Legislative and regulatory aspects

Strengthening critical infrastructure protection has been one of the Obama administration's objectives. Executive Decree 13636 of February 2013 put in place a number of measures. First, the Department of Homeland Security (DHS) and the Department of Defense (DOD) have been tasked with establishing procedures to begin sharing cybersecurity information with owners and operators of critical infrastructure. It identifies 16 areas of critical infrastructure and asks the NIST to propose a framework for controlling these risks. In February 2014, the NIST published its framework (section 6.3.1). Based on these elements, the Cybersecurity Act was signed on December 18, 2015 by President Obama. It contains the text on information sharing (Cybersecurity Information Sharing Act). The purpose of the legislation is to promote and encourage the private sector and the U.S. government to exchange information on cyber threats quickly and responsibly. Under the law, information about a threat found on a system can be quickly shared in order to prevent a similar attack or mitigate a similar threat to other companies, agencies and consumers.

Cybersecurity legislation has also been strengthened in Europe in recent years, and regulatory compliance is an important issue for companies. The threat of severe financial penalties for non-compliance with established rules is real.

In addition to regulations applicable to all information systems, industrial installations may be affected, as they belong to an operator of vital importance, and fall under the Military Programming Act (LPM) (Legifrance 2018b) replaced by the NIS (EU) 2016/1148 directive of July 6, 2016 (Europe NSI 2016). They may also be part of the Seveso classified establishments and be covered by Directive 2012/18/EU of July 4, 2012, known as the Seveso 3 Directive (Europe Seveso III 2012).

The first category of facilities is that for which a cyber-attack can lead to an inability to produce or provide the required services and disrupt a country's economic and social life. The second category is that for which a cyber-attack can cause an industrial accident and have an impact on populations.

The NIS Directive has been transposed into French law. The law (France Loi no. 2018-133 2018) was adopted on February 15, 2018 and promulgated on February 26, 2018. Its objective is to define measures to ensure a high level of network and information system security. It aims to protect against cyber-attacks against certain strategic companies. It introduces two new categories of actors, which will be subject to higher standards of IT security:

– operators of essential services (OES);

– digital service providers (DSP).

In France, it is in line with the Military Programming Act of 2013, the first step in the implementation of the new strategic guidelines of the White Paper on Defense and National Security. More than 200 companies, operating facilities or using facilities and structures whose unavailability would have a significant impact on the country's security and functioning, are classified as "operators of vital importance (OVIs)". They are divided into 12 business sectors. The law requires them to strengthen the security of the critical information systems they operate, which are called information systems of vital importance.

A business sector of vital importance (BSVI), as defined by Article R. 1332-2 of the French Defense Code, consists of activities contributing to the same objective, which:

– relate to the production and distribution of essential goods or services (where these activities are difficult to substitute or replace): satisfaction of basic needs for the life of populations, exercise of State authority, functioning of the economy, maintenance of defense potential, national security;

– can present a serious danger to the population.

An OVI, as defined by Article R. 1332-1 of the Defense Code, is an organization that:

– carries out activities included in a vitally important sector of activity;

– manages or uses for this activity one or more establishments or works, one or more installations whose damage, unavailability or destruction as a result of an act of malicious intent, sabotage or terrorism, would directly or indirectly risk seriously jeopardizing the Nation's combat or economic potential, its security or capacity to survive.

The security rules to be respected are presented for each BSVI and are classified by theme (France Annex JORF 2016). For OESs (or OVIs in France), the main obligations of the law transposing the NIS Directive are as follows:

– a declaration to ANSSI, without delay after becoming aware of it, of incidents affecting the networks and information systems necessary for the provision of essential services, where such incidents have or are likely to have, taking into account in particular the number of users and the geographical area affected as well as the duration of the incident, a significant impact on the continuity of these services;

– an obligation to identify the risks that threaten the security of these networks and information systems, and an obligation to take the necessary and proportionate technical and organizational measures to manage these risks, in the following areas:

– security of systems and installations;

– incident management;

– business continuity management;

– monitoring, audit and control;

– compliance with international standards:

– cooperation during ANSSI controls.

Sanctions can be significant:

– in the event of failure to report incidents, a fine of 75,000 euros;

– in the event of an obstacle to ANSSI controls, a fine of 125,000 euros;

– in the absence of security measures or failure to comply with security rules, a fine of 100,000 euros.

In line with the NIS Directive, the European Council called at the end of 2017 for a common EU approach to cybersecurity. New proposals (European Council 2017) have been made, such as the establishment of an EU cybersecurity agency, giving ENISA (European Union Agency for Network and Information Security) greater powers and establishing an EU-wide cybersecurity certification system.

It should also be mentioned that, for traditional information systems, a very important regulation concerns the management of personal data. These are subject to the General Data Protection Regulation (GDPR) (Europe GDPR 2016), applicable since May 25, 2018. It replaces existing regulations and has been designed to harmonies data privacy laws across Europe and to strengthen the protection of private data. This regulation also applies to industrial systems for their part in data management.

7

The Approach Proposed by Standard 62443

7.1. Presentation

The IEC 62443 standard consists of a family of normative texts aimed at the various cybersecurity actors involved in the lifecycle of an industrial system. Its objective is rather ambitious, as it covers many aspects. The standard was initially developed by the International Society of Automation (ISA) under the name ISA 99. Work began in 2002, the first technical reports were published in 2004 and the first ANSI/ISA standards were published in 2009. This work has integrated the National Institute of Standards and Technology's main ideas. The convergence of ISA's work with that of the IEC began in 2011. Development is delegated to a committee of experts with various experiences and fields of activity. Currently, not all texts are final, but the fundamental concepts are defined. The structure of the standard is given in section 7.3.

The target audience for this standard is the industrial control system (ICS) stakeholder community: asset owners and operators, system integrators, product suppliers, service providers, and even government agencies and regulatory bodies with the legal authority to conduct audits to ensure compliance with applicable laws and regulations. Each of them will use the standard on a part of the equipment or at a specific stage of the lifecycle, in particular:

– operators or owners of assets, to carry out a management of the security and express security requirements for the different parts of the installation;

– system integrators, product suppliers and service providers to assess whether their products and services can provide the level of security needed to meet the security requirements required by the asset owner.

The requirements and guidelines of the IEC 62443 standard are defined for persons who have responsibilities related to the implementation, operation, maintenance or decommissioning of industrial automation and control systems. This includes operators, product suppliers, system integrators, system users and suppliers.

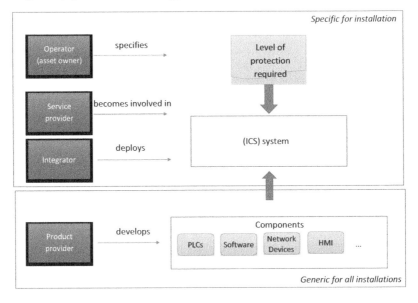

Figure 7.1. *The target audience for the standard*

The scope of application concerns control systems for industrial installations in the broad sense: systems with automation, supervision systems (SCADA) for industrial processes, but also control systems for transport or energy networks, as well as installations including IIoT equipment. The standard concerns industrial IT systems and is complementary to the ISO 27000 family used for non-industrial IT systems. It has evolved to organize itself in a way that is compatible with the latter, in particular with regard to the security management system or the structure of the catalogue of measures.

Having been designed by functional safety practitioners, it shares a common philosophy with IEC 61508 (Chapter 8), in particular with regard to the concept of the level of safety to be expected to guarantee a given level of risk control, taking into account the risk issues.

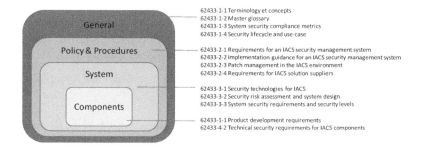

Figure 7.2. *The different aspects covered by the standard*

7.2. IACS lifecycle and security stakeholders

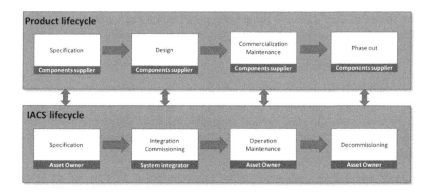

Figure 7.3. *IACS lifecycle*

In the context of IEC 62443, ICSs are called IACS (Industrial Automation and Control System). Mastering cybersecurity in an IACS requires the involvement of all relevant stakeholders, who share "a responsibility" for all phases of the cybersecurity lifecycle:

– product suppliers must safely develop commercial components (COTS) that include adequate security capabilities to achieve the desired level of security at the system level;

– system integrators must use practices that result in secure solutions that support cybersecurity requirements for the planned deployment environment;

– asset owners must configure, commission, operate and maintain the deployed solution in accordance with the documentation, so that the solution's cybersecurity capabilities do not degrade over time;

– service providers must also implement practices that comply with cybersecurity requirements.

The standard contains elements for each of these stakeholders, as detailed below.

7.3. Structure of the IEC 62443 standard

IEC 62443 is composed of several parts (ISA 2018; Hauet n.d. a, 2012). Its structure is shown in Figure 7.4. The grayed elements are those that define the standard, the others are technical reports. Some parts continue to evolve. The standard is divided into several parts that are articulated on four levels.

The first level, entitled "General", consists of four parts:

– 62443-1-1 is a standard that introduces the concepts and models used in the series. The target audience includes all those who wish to familiarize themselves with the fundamental concepts that form the basis of the series. The March 2017 version has been enriched and presents the important concepts in detail;

– 62443-1-2 is a technical report that provides a glossary of terms and abbreviations used throughout the series;

– 62443-1-3 is a document that describes a series of quantitative metrics derived from the fundamental requirements (FRs). This document is currently being drafted;

– 62443-1-4 is a technical report that provides a detailed description of the security lifecycle of IACS, as well as several examples of use.

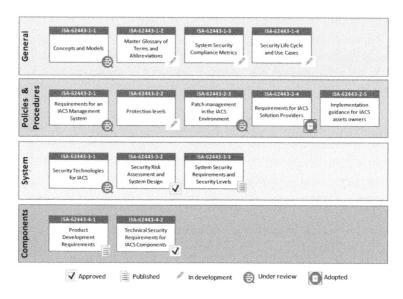

Figure 7.4. *Structure of the IEC 62443 (7/2018) standard. For a color version of this figure, see www.iste.co.uk/flaus/cybersecurity.zip*

The second level concerns the organizational aspects described by policies and procedures:

– 62443-2-1 is a standard that describes what is required to define and implement an effective IACS information security management system. The target audience includes end users and asset owners who are responsible for the design and implementation of such a program;

– 62443-2-2 is a guide for assessing the level of protection against cybersecurity threats of an operational IACS;

– 62443-2-3 is a document on patch management for IACS. The target audience includes all those responsible for the design and implementation of a patch management approach;

– 62443-2-4 is a standard that specifies requirements for IACS providers. The main audience includes suppliers of control system solutions. This standard was developed by the IEC TC65 GT10 group and adopted by the ISA;

– 62443-2-5 is a section providing guidance on what is required to operate a cybersecurity management system for an IACS.

The third level concerns the requirements at the system level:

– 62443-3-1 is a technical report that describes the application of various security technologies (authentication system, firewall, encryption, etc.) in an IACS environment. The target audience includes all those who wish to learn more about the applicability of specific technologies in a control system environment;

– 62443-3-2 is a standard that describes risk assessment methods and the design of IACS. This standard is mainly intended for asset owners or end users;

– 62443-3-3 is a standard that describes the fundamental security requirements (FRs) and security levels (SLs).

The fourth level concerns the requirements at component level:

– 62443-4-1 is a standard that describes the requirements for product development. The main audience includes suppliers of control system solutions;

– 62443-4-2 is a standard that contains a set of requirements derived from the system requirements (SR and RE of 62443-3-3) for components. The main audience includes suppliers of control system solutions.

7.4. General idea of the proposed approach

The IEC 62443 standard is a set of voluminous documents covering many aspects of ICS cybersecurity. In particular, it proposes:

– the definition of a management system (ISMS) that is aligned with ISO 27001 and 27002 and includes a list of requirements and measures;

– the specification of an approach for risk analysis;

– the notion of a system's and component's security level;

– the notion of FRs to be met to reach this level;

– the notion of a zone and conduit;

– the notion of a lifecycle;

– the explicit definition of the various stakeholders involved in the design, construction or operation of an installation: asset owner, system integrator, component supplier;

– a catalogue of measures based on the NIST 800-53 standard with adaptations for ICS.

The objective of this standard is to provide a framework for implementing a cybersecurity management system for an industrial facility that covers the facility's entire lifecycle. Chapter 3 of this book presents the general outline of an ISMS, a key step of which is the risk analysis step.

The analysis approach proposed by IEC 62443 consists of two main phases (Figure 7.5):

– Phase 1: a high-level global risk analysis whose objective is to break the main system down into homogeneous zones in terms of risk, linked by conduits. These terms are defined in the following;

– Phase 2: a detailed analysis of each zone that needs it in order to determine the level of security (SL) required depending on the level of risk, and then define the measures to achieve it.

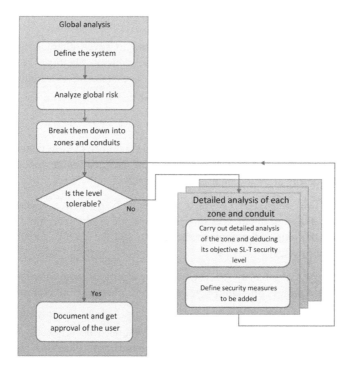

Figure 7.5. *Risk analysis according to IEC 62443*

The idea of defining an SL comes from the functional safety approach that defines a SIL level (Chapter 8). For cybersecurity, the scalar magnitude is replaced by a vector, as several aspects must be considered. These aspects are called FRs and are seven in number.

IEC 62443-3-3 at system level and IEC 62443-2 at component level describe a list of security requirements that must be met to achieve a given level for a FR.

The risk analysis makes it possible to establish a list of measures to be implemented and a security policy. It can also be the first step of the security management system (ISMS), which is described in volume IEC 62443-2-1. This document organizes the process according to ISO27001 and 11 categories from ISO 27002, which are adapted to the case of IACS.

7.5. Basics of the standard

7.5.1. *Fundamental requirements*

The fundamental requirements correspond to seven characteristics of a system or of a component characterizing the security. They can be satisfied with a relatively high security level (SL). The standard distinguishes five levels, from 0 to 4, presented in the following section.

The seven FRs are as follows:

– FR1: identification and authentication control (IAC) measures the ability to identify and authenticate all human users, software processes and equipment before allowing them to access IACS;

– FR2: use control (UC) characterizes the level at which it is ensured that all identified users (human, software processes and equipment) have privilege to perform the required actions on the system and the level of monitoring on use of these privileges;

– FR3: system integrity (SI) measures the ability to ensure the integrity of information (protection against unauthorized changes) in communication channels and data repositories;

– FR4: data confidentiality (DC) measures the capacity of non-dissemination of information in communication channels and data repositories;

– FR5: restrict data flow (RDF) characterizes the level of segmentation of the system into zones and conduits to avoid unnecessary data propagation;

– FR6: time response to an event (TRE) measures the level of ICS operational monitoring and the ability to respond to incidents in a timely manner;

– FR7: resource availability (TA) measures the level of protection to ensure system availability against denial of service attacks, the ability to operate in degraded mode and the ability to recover from them.

To assess the security level of a component or system for one of the seven FRs, the standard provides a list of criteria to be validated. A subset of the criteria is required to obtain level 1, 2 or 3. The higher the level, the more criteria have to be met.

The criteria for system-level analysis are provided in IEC 62443-3-3. An extract is presented in Tables 7.1 and 7.2. For example, if SR1.1 is satisfied for networks without trust, we have a level 3; if this same criterion is satisfied for all networks, we have a level 4. The score assigned to the analyzed element will be the lowest obtained among all the criteria. The complete table can be found in Appendix 5.

	SL1	SL2	SL3	SL4
FR 1 – Identification and authentication control (IAC)				
SR 1.1 – Human user identification and authentication	x	x	x	x
SR 1.1 RE 1 – Unique identification and authentication			x	x
SR 1.1 RE 2 – Multifactor authentication for untrusted networks			x	x
SR 1.1 RE 3 – Multifactor authentication for all networks				x
SR 1.2 – Software process and device identification and authentication		x	x	x
SR 1.2 RE 1 – Unique identification and authentication			x	x
SR 1.3 – Account management	x	x	x	x
SR 1.3 RE 1 – Unified account management			x	x
SR 1.4 – Identifier management	x	x	x	x
SR 1.5 – Authenticator management	x	x	x	x
SR 1.5 RE 1 – Hardware security for software process identity credentials			x	x
SR 1.6 – Wireless access management	x	x	x	x
SR 1.6 RE 1 – Unique identification and authentication			x	x

Table 7.1. *SL level evaluation grid for systems*

For component level analysis, a similar approach exists. The requirements are classified into requirements applicable to all types of components and into requirements specific to each family:

– applications;

– embedded devices;

– host devices;

– network devices.

	SL1	SL2	SL3	SL4
FR 1 – Identification and authentication control (IAC)				
CR 1.1 – Human user identification and authentication	x	x	x	x
CR 1.1 RE 1 – Unique identification and authentication			x	x
CR 1.1 RE 2 – Multifactor authentication for all interfaces				x
CR 1.2 – Software process and device identification and authentication		x	x	x
CR 1.2 RE 1 – Unique identification and authentication			x	x
CR 1.3 – Account management	x	x	x	x
CR 1.4 – Identifier management	x	x	x	x
CR 1.5 – Authenticator management	x	x	x	x
CR 1.5 RE 1 – Hardware security for authenticators			x	x
CR 1.6 – Wireless access management	x	x	x	x

Table 7.2. *SL level evaluation grid for components*

7.5.2. Security Levels (SL)

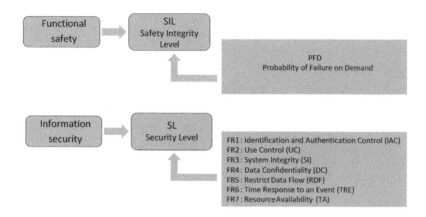

Figure 7.6. *Comparison of SIL and SL*

The security levels (SL) represent the confidence that can be placed in the ability of a system, zone and/or its components to provide a certain degree of security in relation to the fundamental requirements (FR).

As mentioned above, the idea is to propose the equivalent of the SIL level (Chapter 8) for cybersecurity. This indicator, called SL, has several components, one for each FR. Each component can have a value from 0 to 4 (Figure 7.7). A component or system will be characterized by an SL level, which can be written as {2 1 2 1 1 1 0 2}. This means that the level for FR1 is 2, for FR2 is 1, etc. The SL may concern an equipment, the software, a zone or any element or group of elements belonging to the installation. The SL of an element, such as a PLC, or a zone, is noted as follows:

SL(PLC1) = { 1 1 2 1 0 0 1 } or SL(Zone10) = { 1 1 2 1 0 0 1 }

The standard defines the security level to be achieved SL-T (target), the level achieved SL-A (achieved) and the possible level SL-C (capability):

– the risk analysis of the installation makes it possible to determine the level of cybersecurity required for each zone and each conduit. For example, a facility that can cause significant damage to the population and uses an SIS is at high risk. In this case, the required security level, written as SL-T (target), will be level 4;

– the SL-A level is the level actually reached by the automated system. A list of approximately 100 technical criteria is used to quantify the SL of a system for each of the components corresponding to a FR. Countermeasures are determined in such a way that the SL of the installation with these countermeasures achieves the set SL-T objective. This analysis is carried out by zones;

– the SL-C is used to characterize a component (hardware or software) and represents the level of security that can be achieved if the component is implemented correctly.

The general format of the SL vector is as follows:

SLx([FR,]domain) = {IAC UC SI DC RDF TRE RA},

where x = T, A or C; FR is a fundamental requirement and "domain" characterizes a component or zone. For example:

SL-T(zone1) = {2 2 0 1 3 1 3}

SL-C(RA, PLC) = 4

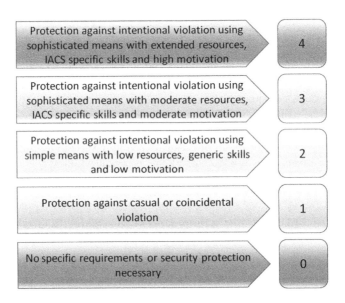

Figure 7.7. *Security levels. For a color version of this figure, see www.iste.co.uk/flaus/cybersecurity.zip*

Each level has an explicit meaning:

– SL0: no protection. The SL0 level has several meanings depending on the situation in which it is used. For SL-C, this means that the component or system does not meet some of SL1's requirements for this particular FR. In the definition of SL-T for a particular zone, this means that the results of the risk analysis indicate that no requirements are required for a given FR, given the level of risk. For SL-A, this means that the zone under consideration does not meet some of SL1's requirements for this particular FR;

– SL1: protection against usual or coincidental violations. Occasional or coincidental security violations are generally due to lax enforcement of security policies. These can be caused by well-intentioned employees or by an external threat. Many of these violations will be related to the security program, and addressed by strengthening the application of policies and procedures. An example of such a violation is the modification of a setpoint by an operator outside the authorized areas during normal operation, because the appropriate use restrictions are not implemented;

– SL2: protection against intentional violation by using simple means with low resources, generic skills and low motivation. Simple means do not require much knowledge from the attacker. The attacker does not need a detailed knowledge of the security, domain or particular system being attacked. These attack vectors are well known, and there may be automated tools to help the attacker. They are also designed to attack a wide range of systems instead of targeting a specific system, so that an attacker does not need a significant level of motivation or resources. An example could be a virus that infects a workstation in the demilitarized zone (DMZ), where a USB key is introduced and then spreads over the network. Another example could be an operator surfing on a website on the HMI located in the industrial network and downloading a Trojan horse. Attacks of this level can therefore be of a frightening effectiveness;

– SL3: protection against intentional violation using sophisticated means with moderate resources, specific IACS skills and moderate motivation. Sophisticated means require advanced security knowledge, advanced domain knowledge, advanced knowledge of the target system or any combination of these elements. The system is specifically targeted. The attacker may use exploits in unfamiliar operating systems, weaknesses in industrial protocols, specific information about a particular target to violate system security, or other means requiring more motivation, skills and knowledge than necessary for SL1 or SL2. An example would be an attacker who accesses the data historian server using a brute force attack through the industrial/corporate DMZ firewall from the company's wireless network;

– SL4: protection against intentional violation by using sophisticated means with extensive resources, specific IACS skills and strong motivation. SL3 and SL4 are very similar in that they both involve sophisticated means used to violate the system's security requirements. The difference comes from the fact that the attacker is even more motivated and has extensive resources at his disposal. An example would be a crime organization with the motivation and resources to develop or purchase customized zero day exploits.

7.5.3. Zones and conduits

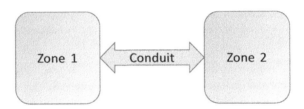

Figure 7.8. *Zones and conduits*

Very often, and especially for complex ICS, it is not necessary to require the same level of security for all elements. To manage these different levels, the standard proposes to break the installation down into zones. These zones are connected by conduits. The level of security required for a zone will be provided by the risk analysis and the conduits are secured in such a way as to allow the coexistence of zones of different levels.

A zone is defined as a logical or physical grouping of assets that share common security requirements. A zone has a boundary that separates the elements within the zone from those outside. Information and staff move within and between zones. From these exchanges, it is possible to define accesses that may be electronic communication channels or physical access for people.

A zone can be defined according to physical or logical criteria (virtual zone). In the first case, elements are grouped according to their location. In the second, for virtual zones, functional or other criteria are used to group them together.

A zone can be divided into subzones, which defines security layers and allows for defense-in-depth.

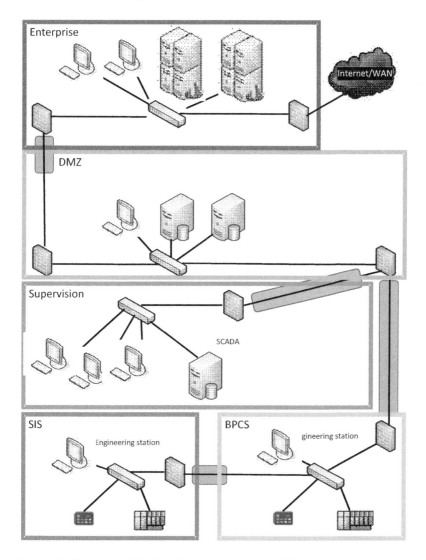

Figure 7.9. *Example of division into zones and conduits. For a color version of this figure, see www.iste.co.uk/flaus/cybersecurity.zip*

A conduit is defined as a particular zone that groups together the elements that allow communication between two zones. It can be seen as a path for the information flow between two zones. It can provide security functions to allow secure communication. Any electronic data transfer between two zones must be done via a conduit. A zone or conduit may or may not be trusted.

To define a zone, in addition to its identifier, it is necessary to give its physical or logical boundary, the list of access points, the data flows at each access point, the conduits and connected zones, and the list of assets.

A conduit is defined by its border, its end points (physical equipment and applications) that use communication channels, the data flows it carries, the connected zone and its assets.

An example of zone partitioning is shown in Figure 7.9.

7.5.4. *Maturity level*

The IEC 62443 standard defines maturity levels used to characterize organizational measures. These are defined according to the CMMI-SVC model defined in CMMI (2010). It has been simplified and has only four levels.

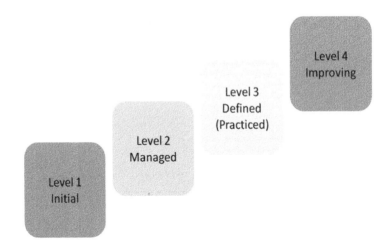

Figure 7.10. *Maturity levels. For a color version of this figure, see www.iste.co.uk/flaus/cybersecurity.zip*

Level	Classification	Meaning
1	Initial	Processes are carried out on *an ad hoc basis* and often not documented (or not fully documented). Therefore, consistency over time is not guaranteed. NOTE: "Documented" in this context refers to the procedure followed in carrying out this activity (e.g. detailed instructions to staff), not the results of the process execution. In most IACS, changes resulting from the execution of a procedural process are often documented.
2	Managed	There is documentation that describes how to manage the achievement and performance level of the capacity. This documentation may take the form of written procedures or written training programs for capacity execution. There may be a delay between the definition of the process and its repeated implementation. The discipline reflected by maturity level 2 ensures that practices are reproducible, even in times of stress. Where these practices are in place, their implementation will be carried out and managed in accordance with their documented plans.
3	Defined	A level 3 process is a level 2 process that is practiced throughout IACS. The performance of a level 3 service can be shown as repeatable over time within the IACS.
4	Improving	At this level, the CMMI-SVC levels "Quantitatively Managed" and "Optimizing" are combined. By using appropriate key process performance indicators, process efficiency and/or performance improvements can be demonstrated. This results in a security program that improves the process through technological/procedural/management changes.

Table 7.3. *Maturity levels*

7.5.5. *Protection level*

The standard defines the notion of level of protection, which is obtained as a combination of the technical security level and the level of maturity of the associated organizational measures. The assumption is that technological measures make it possible to define the maximum achievable level of protection, but that this can only be achieved if these technical measures are implemented in a managed process.

An example of a matrix for assessing the level of protection is shown in Figure 7.11.

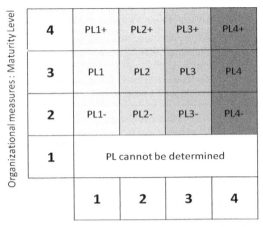

Figure 7.11. *Maturity levels*

7.6. Risk analysis

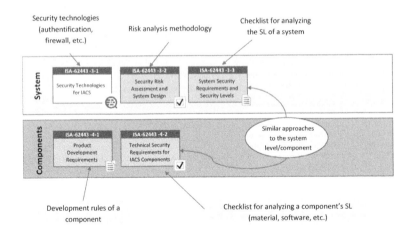

Figure 7.12. *Component and system level documents. For a color version of this figure, see www.iste.co.uk/flaus/cybersecurity.zip*

7.6.1. *General approach*

The implementation of IEC 62443 begins with a risk analysis. The approach is detailed in document 62443-3-2 (Figure 7.12). The steps are as follows:

– ZCR1: identify the SuC (System under Consideration);

– ZCR2: carry out a high-level risk analysis;

– ZCR3: partition into zones and conduits;

– ZCR 4: if the overall risk level exceeds the tolerable level;

– ZCR5: carry out a risk analysis of each zone;

– ZCR6: document the requirements for cybersecurity;

– ZCR7: obtain approval from the owner of the asset.

This process begins with the identification of the system under consideration, which involves defining its scope and access points. The result of this definition is a list of assets and can be represented by an architecture diagram.

The second step consists of carrying out a global risk analysis of the SuC. The objective is to identify the worst case, as well as the risk generated by a malfunction of the IACS. It is of course possible to rely on industrial risk or functional safety analyses (Chapter 8). The level of risk is assessed with a risk matrix and allows us to situate it in relation to what is tolerated by the organization.

The ZCR3 step consists of partitioning the SuC into zones and conduits. This breakdown is done in accordance with the definitions in section 7.5.3, the objective being to prepare the detailed analysis. We are therefore aiming to obtaina a given level of security for the zone. Important rules are:

– to separate the IT zone from the OT zone(s);

– to define specific zones for the SIS;

– to define specific zones for temporarily connected equipment;

– to define zones for wireless networks;

– to separate the zones connected via external networks.

Zones and conduits must be documented and can be represented on an architecture diagram.

Step ZCR4 is to determine if the overall risk level exceeds the tolerable level. If the answer is yes, a detailed analysis of each zone is performed in step ZCR5. The result of this analysis is, for each zone and conduit, an SL-T target security level, defined according to the risk level of the zone or conduit in question. This step is detailed in section 7.6.2.

Step ZCR6 is devoted to writing the specifications for cybersecurity requirements. These contain:

– a description of the SuC, with at least a synthetic description of its function and the process or controlled equipment;

– a description of the physical and logical environment in which IACS is located;

– a description of the threats and sources of threats identified;

– mandatory, technical and organizational security measures;

– the acceptable level of risk;

– where applicable, the regulatory obligations to which the installation must comply.

The last step ZCR7 consists of obtaining approval of the risk analysis by the persons in charge of IACS responsible for the security, integrity and reliability of the process controlled by the SuC.

7.6.2. *Detailed risk analysis*

This analysis is carried out on a zone or a conduit by the following steps:

– identify threats that affect the zone or conduit. More details are given in Chapter 4;

– identify vulnerabilities. The zone or conduit is analyzed in detail to identify vulnerabilities and document them. This analysis can be done using various sources of documentation or vulnerability analysis tools (Chapter 5, section 5.7);

– determine the consequences and impacts. The consequences must be assessed by considering the most unfavorable impact in different zones such as staff security, financial losses, business interruptions and environmental impacts;

– assess the likelihood of threats without taking countermeasures into account. Most often this evaluation is qualitative;

– determine the level of risk associated with cybersecurity without countermeasures. To do this, a risk matrix is used;

– determine the target security level SL-T. This level is determined from the level of risk of the zone;

– identify and evaluate existing countermeasures. These must be identified and evaluated to determine their effectiveness in reducing likelihood or impact;

– reassess likelihood and severity with countermeasures;

– determine the residual risk. The residual risk identified for each threat identified in step 1 is assessed;

– compare the residual risk with the tolerable risk. If the residual risk exceeds the tolerable risk, the organization shall determine whether the residual risk is accepted, transferred or reduced, following the general approach described in Chapter 3;

– identify additional countermeasures. Measures must be proposed to reduce the risk, unless it was decided at the previous step to accept or transfer it;

– document the analysis and communicate the results to stakeholders.

7.6.3. *Determination of SL-T*

The approach of the standard 62443 is to define SL-T according to the level of risk. Several methods can be used. The simplest way is to assign an SL-T level to each risk level (Figure 7.13).

IEC 62443-3-2 also proposes an approach based on calculation of the Cyber Risk Reduction Factor (CRRF). The level of risk is calculated as the product of likelihood and impact $R = P \times I$. The CRRF is defined as follows: CRRF= R/Rtol, where R is the risk level and Rtol is the tolerable level. The level of security is then calculated by the relationship:

$$\text{SL-T} = min\left\{4, \left(CRRF - \frac{1}{Rtol}\right)\right\}$$

For a tolerable risk level of 4, we obtain the table given in Figure 7.14. However, this approach should be used with caution (Braband 2017).

7.6.4. Countermeasures

When the SL-T, target level, is determined, the next step is to analyze the zone using Table 7.1, or the one given in Appendix 5. The basic levels SR define the security measures to be taken to achieve a level of at least 1, and to increase this level, Requirement enhancements (RE) are required. SR and RE can be found in the table defining the FR in Appendix 5.

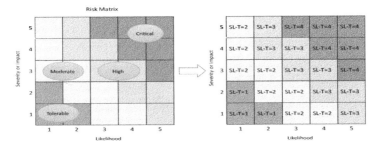

Figure 7.13. *Determination of SL-T from the risk matrix. For a color version of this figure, see www.iste.co.uk/flaus/cybersecurity.zip*

Tolerable Risk=4			Tolerable Risk=4		
Risk	CRRF	SL-T	Risk	CRRF	SL-T
1	0.25	0	16	4	3
2	0.5	0	17	4.25	4
3	0.75	0	18	4.5	4
4	1	0	19	4.75	4
5	1.25	1	20	5	4
6	1.5	1	21	5.25	4
7	1.75	1	22	5.5	4
8	2	1	23	5.75	4
9	2.25	2	24	6	4
10	2.5	2	25	6.25	4
11	2.75	2			
12	3	2			
13	3.25	3			
14	3.5	3			
15	3.75	3			

Figure 7.14. *Determination of SL-T with CRRF. For a color version of this figure, see www.iste.co.uk/flaus/cybersecurity.zip*

The general approach to implementing security measures is presented in Chapter 11.

7.7. Security management

Document 62443-2-1 defines requirements for the establishment, implementation, maintenance and continuous improvement of a security program for an IACS. These requirements, when applied conscientiously, provide security capabilities aimed at reducing risks. These requirements are formulated to be independent of implementation, allowing asset owners to choose the approaches best suited to their needs.

As an IACS is a combination of personnel and digital technology, the security program is composed of:

– organizational measures, which include an information security management system and generally define enterprise-wide security policies and practices for computer systems;

– technical security measures implemented by the hardware and software;

– security-related processes, often used to configure and maintain technical security capabilities.

The aspects covered in the standard complete the information security management system of an Information System by taking into account the specificities of IACS. A typical example is the management of updates which, in a traditional system, is made automatically, which is not appropriate for an IACS, as this could cause malfunctioning. Specific measures are therefore proposed.

The standard proposes a number of requirements grouped into domains, called Security Program Elements (SPE):

– SPE1 – organizational security;

– SPE 2 – configuration management;

– SPE 3 – network security;

– SPE 4 – component security;

– SPE 5 – protection of data;

– SPE 6 – user security;

– SPE 7 – event and incident management;

– SPE 8 – system integrity and availability.

Requirement classes, called Security Control Classes (SCC), are associated with domains. For example, for SPE1, three classes are defined:

– SCC 1: security management system;

– SCC 2: security assessment and review;

– SCC 3: security of physical access.

Then, for each class, requirements are defined based on those of ISO 27001 for generic requirements, or specifically if they are specific to IACS.

For example, for ORG 3 (Table 7.4), the requirements are those of ISO 27001, except for the one concerning access to IACS, which is not an issue for a conventional computer system.

Specific	ORG 3.1	Physical access to the IACS, including access to facilities, equipment and cabling, shall be controlled to meet acceptable risk targets
ISO 27000	A.11.1.1	Physical security perimeter
	A.11.1.2	Physical entry controls
	A.11.1.3	Securing offices, rooms and facilities
	A.11.1.4	Protecting against external and environmental threats
	A.11.1.5	Working in secure areas
	A.11.1.6	Delivery and loading areas
	A.11.2.1	Equipment siting and protection
	A.11.2.3	Cabling security
	A.11.2.8	Unattended user equipment
	A.12.1.4	Separation of development, testing and operational environments

Table 7.4. *Requirements for ORG 3*

7.8. Assessment of the level of protection

Part IEC 62443-2-2 of the standard defines a set of requirements to be met for each class of security measures (SCC). These are both organizational and technical. For an existing facility, they can be evaluated based on the observation of the IACS and its organizational functioning. Organizational measures are assessed on the maturity scale, from 1 to 4, while technical measures are assessed on the security level scale, from 1 to 4. For each class, a list of associated requirements is built:

– from standard 62443-2-1 for organizational requirements for ISMS implementation;

– from standard 62433-2-4 for service provider requirements;

– from standard 62443-3-2 for risk analysis requirements;

– from standard 62443-3-3 for technical system requirements;

– from standard 62443-4-1 for product development requirements;

– from standard 62433-4-2 for technical requirements at component level.

The evaluations obtained are used to determine the level of protection as detailed in section 7.5.5.

7.9. Implementation of the IEC 62443 standard

IEC 62443 is a multifaceted standard for all stakeholders in industrial cybersecurity. It can be used in different ways.

7.9.1. Certification

Two documents are addressed to product suppliers: IEC 62443-4-1 (product development requirements) covers all requirements on product development processes, and IEC 62443-4-2 (technical security requirements for IACS components) defines CR requirements for components, derived from system requirements (SR, RE). These documents can be used as a basis for component certification by ISASecure[1] or TÜV[2], for example.

Components certified to a certain SL level (SL-C) allow the security level assessment to be carried out in a rigorous manner (section 7.5.2).

---

1 Available at: www.isasecure.org.
2 Available at: https://www.tuv.com/.

7.9.2. *Service providers and integrators*

Most ICSs are developed and installed by integrators. System maintenance is carried out by external parties. These interventions are a very significant source of threat and the risk is high.

IEC 62443-2-4 (requirements for IACS solutions suppliers) focuses on certification of IACS suppliers' security policies and practices. This document has a status "adopted" in the standard.

7.9.3. *IACS Operators*

The development of a secure solution and the implementation of a cybersecurity management system must follow a rigorous and proven methodology. The IEC 62443 standard proposes one. It is based on concepts such as the division into zones and conduits, risk analysis (global and detailed) and the implementation of security measures. This standard is not isolated from other approaches and, over the years, there has been some convergence, particularly with the ISO 27000 family, on the common points of the two approaches. The approach described in this book (Chapter 11) follows the general approach of standard 62443.

8

Functional Safety and Cybersecurity

8.1. Introduction

8.1.1. *Components of operational safety*

Cybersecurity issues, and more specifically the cybersecurity of industrial and cyber-physical systems, are relatively new. They have been of concern to designers for less than 20 years for the reasons presented in the Introduction.

On the other hand, ensuring that a system operates reliably and without generating risks for those around it, is an older issue. Approaches to improve system reliability were developed from the Second World War and in the 1960s, during the space conquest. Initially, the causes of failures were mainly mechanical, but then the electronic and programmable electronic aspects became more and more important. Practices were standardized in the 2000s with the publication of standard IEC 61508 (2010). The set of methods used to ensure the proper functioning of a system is part of a discipline called "operational safety" (Figure 8.1), in which we consider the availability (an entity's ability to provide the required service at a given time) and safety (the requirement not to harm people, the environment, or any other assets during a system's lifecycle).

These two properties are themselves based on the reliability of the system under consideration. This notion is defined as the ability of an entity to perform a required function, under given conditions, over a given time interval and is measured by the probability that an entity will operate over the interval $[0,t]$.

194   Cybersecurity of Industrial Systems

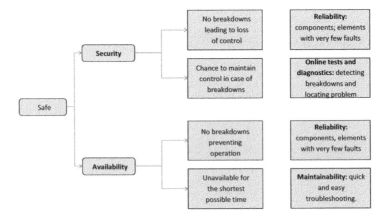

Figure 8.1. *Components of operational safety (RAMS: Reliability, Availability, Maintainability, and Safety)*

The steps of a functional or operational safety study are aligned with those of the traditional risk management process:

– definition of the context: what is the system analyzed, who are the stakeholders, operating constraints, regulatory obligations;

– risk analysis;

– definition of safety requirements;

– implementation of measures (allocation of SIL levels, concept detailed below).

The specific points of operational safety concern how to measure risks and how to quantify protection systems with a Safety Integrity Level (SIL) in order to reach a given level of risk.

The approaches used in operational safety and cybersecurity are similar and must be conducted jointly. Indeed, to ensure the overall proper functioning of a system, it appears that the risk analysis carried out in relation to the different types of hazards (technical, human or organizational) must be supplemented by an analysis of the risks associated with cyber-attacks for elements containing programmed devices.

Consider, for example, a chemical manufacturing process where a control loop is used to control the temperature by regulating the flow rate of a cooling fluid. An analysis of the causes of the malfunction makes it possible to identify:

– a measurement problem related to the sensor, which may be due either to a hardware failure of the sensor or to incorrect calibration, itself being caused by human error or being intentional and due to a cyber-attack;

– a problem related to the cooling valve, which may be due either to a hardware failure or an order error, itself being caused by a hazard or a cyber-attack;

– a change in the temperature set point, which can also be caused by human error or a cyber-attack.

Suppose that a loss of control leads to an explosion, an analysis of the consequences will identify:

– damage to health for staff and surrounding populations;

– production losses and associated financial losses;

– environmental damage;

– violation of regulatory provisions;

– material damage.

In this example, it may be noted that the causes of the feared event are distinct and complementary, but the consequences are common.

This is the case most of the time; cybersecurity and operational safety analyses have many similarities. However, there are differences between the two points of view (Table 8.1) that must be taken into account in a comprehensive study. Solutions are not always easy to determine: for example, how do we decide how often to update software that reduces IT vulnerabilities, but may increase operational risks?

The rest of the chapter presents important concepts in the operational safety approach, such as the SIL, the main lines of standard IES 61508 and the main risk analysis methods. This will be useful for positioning them in relation to the methods used in cybersecurity and for proposing a coordinated approach (Chapter 9).

Characteristics	Operational safety	Cybersecurity
Nature of risk	The risk is due to a random, rather rare failure of an element	The risk is due to a deliberate attack, systematically exploiting one or more vulnerabilities with intent
	Unlikely simultaneous failures (except common causes)	The attacker tries to exploit several vulnerabilities simultaneously
	Remote attack nearly impossible	Remote attack very common
Method of analysis	Identification of a feared event, then analysis of the consequences and causes. Methods: FMECA, PHA, fault tree analysis	Identification of security requirements and feared events, analysis of impacts and threats. Method: EBIOS, OCTAVE, CORAS, attack tree, etc.
Example of safety/ security	Safety barriers, redundancy, reduction of common modes, training, work organization	Flow filtering (firewall, diode), secure architecture, IDS, user training, role definition and authentication
Measurement of risk level	Likelihood or probability (usually low) and severity of consequences	Likelihood (often high for the use of known vulnerabilities) and severity of consequences
Specification of level of risk tolerated	SIL (index expressed by a quantitative value)	Vector of level of SL (qualitative)
Example of compatible	Security barriers, redundancy, reduction of common modes, training, work organization	Flow filtering (firewall, diode), secure architecture, IDS, user training, role definition and authentication
Examples of potentially conflicting measures	Infrequent updates to reduce risks	Frequent updates to limit vulnerabilities
	Encryption can be a problem for real time	Encryption improves security

	Fail safe modes, degraded operating modes and emergency stops improve safety	Fail safe modes, degraded operating modes and emergency stops can be used by a cyber attacker to cause the installation to fail
Sharing of experience and Feedback	Many sources of feedback. Future failures are similar to past failures	Feedback must be organized in a global way to secure the discovered vulnerabilities
Allocation of tasks among stakeholders	Strong independence constraint: designer, verifier, etc.	There is no reason why design and validation should not be carried out by the same person
Regulatory aspects	2003 law (2003-699 of July 30, 2003) and Decree 2010 on industrial installations	NIS directive transposed into French law (February 26, 2018) for OIVs
Legal aspects	Liability links can be based on the physical chain (suppliers, maintenance)	It is often difficult to find the source of attacks that may come from a country where legal action is difficult
Certification (references)	A standard is used (IEC 61508) with its sectoral variations	Several competing standards (ISO27000, IEC 62443, CC, CSPN, etc.)

Table 8.1. *Operational security and cybersecurity*

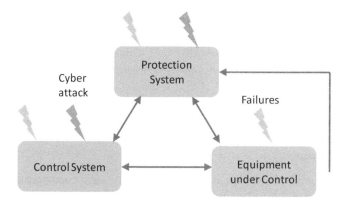

Figure 8.2. *Cyber-attacks and failures*

8.1.2. *SIS and SIL levels*

Among security barriers, a particularly important category is the so-called Safety Instrumented System (SIS). These are automatic systems, designed to react to a feared event. They are made up of (see Figure 8.3):

– one or more sensors;

– one or more processing units (safety controllers, embedded systems);

– one or more actuators (valves or others).

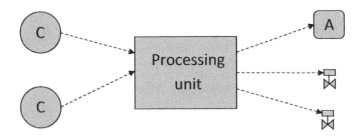

Figure 8.3. *Example of an architecture for an SIS*

A Safety Instrumented Function (SIF) is a function that aims to maintain or bring the equipment under control (EUC) in safety, upon detection of a predefined event or drift. For each safety function, the components of an SIS allow it to implement three subfunctions:

– detection, carried out by a sensor, which is the element allowing the transformation of physical information (pressure, temperature, flow rate, concentration, etc.) into an electrical quantity adapted to the treatment. This value is transformed by a transmitter into a signal, to be transmitted to the processing unit. The transmitted signal can be a 4–20 mA analog signal or an on/off binary signal (digital, or 1/0);

– the treatment, which is more or less complex. It can be limited to a simple display or comparison to generate an actuator command. There are two main technologies, which differ in their level of reliability and processing capacity:

  - wired technologies, based on elementary logic components, such as relays, connected together in an electrically, pneumatically or hydraulically way;

– programmed technologies, based on data acquisition units or alarms, programmable logic controllers, some of which may be dedicated to safety, or industrial computers or microprocessor-based electronic boards, or with programmable logic;

– the action, which is carried out by actuators that transform a signal (electric, pneumatic or hydraulic) into a physical phenomenon that controls a component such as a pump, valve, etc. Depending on the driving energy, we speak of an electric, pneumatic or hydraulic actuator. The actuators are coupled to the end elements: valves, pumps, alarms, etc.

These elements communicate with each other using various technologies: electrical cables, electromagnetic waves, optical fibers, pneumatic or hydraulic pipes.

The more damaging (i.e. more serious) consequences a scenario can generate, the more reliable the safety system to reduce the probability of consequences must be. The security integrity level is a performance measure of the reliability level of this security system.

The Safety Integrity Level (SIL), introduced by IEC 61508 and 61511, is an indicator of the level of functional safety offered by an SIS. There are four levels of SIL (Table 8.2) and two ways to define them, depending on whether the safety system operates in low load mode (such as a protection function) or whether it operates continuously or with high load. Each operating mode has a definition of the probability of failure and a relationship between this magnitude and the level of SIL.

SIL	Request for the SIS		Risk reduction factor
	Uncommon PFD_{avg}	Common PFH	
4	$[10^{-5}, 10^{-4}[$	$[10^{-6}, 10^{-5}[$	10,000–100,000
3	$[10^{-4}, 10^{-3}[$	$[10^{-6}, 10^{-5}[$	10,000–1,000
2	$[10^{-3}, 10^{-2}[$	$[10^{-6}, 10^{-5}[$	1,000–100
1	$[10^{-2}, 10^{-1}[$	$[10^{-6}, 10^{-5}[$	10–100

Table 8.2. *Safety Integrity Level*

SISs are a mean of approaching the tolerable level of risk by performing certain safety functions. There are more and more of them in the systems around us: chemical process emergency shutdown, automatic detection and sprinkling in the event of a fire, gas detection, automatic train shutdown, airbags, driving assistance on cars (ABS), safety systems for medical radiotherapy equipment, etc. In addition, they are also increasingly connected, which poses a major cybersecurity problem. The TRITON attack, presented in Chapter 4, targeted these systems.

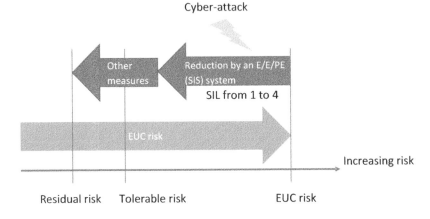

Figure 8.4. *Risk reduction and SIS*

8.2. IEC 61508 standard and its derivatives

A number of standards have been established for the design of the SIS. The general standard is IEC 61508, which was published in 2002. It is entitled "Functional safety of safety-related electrical, electronic and programmable electronic systems", and consists of seven parts that describe how to manage functional safety for equipment using a SIS. These parts are entitled as follows:

– Part 1: "General design requirements (lifecycle)";

– Part 2: "Safety requirements for E/E/PES systems";

– Part 3: "Software requirements";

– Part 4: "Definitions and abbreviations";

– Part 5: "Example of methods for determining safety integrity levels";

– Part 6: "Guidelines for the implementation of parts 2 and 3";

– Part 7: "Presentation of techniques and measures".

Figure 8.5. *Sectoral standards*

This standard has been adapted to different areas:

– process industry: IEC 61511 or ANSI/ISA-84.00.01-2004, functional safety, safety instrumented systems for process production;

– machines: IEC 62061, functional safety of safety-related electrical, electronic and programmable electronic control systems;

– nuclear: IEC 61513, nuclear power plants – instrumentation and control command of systems important to safety;

– railway: IEC 62278/9 and 62425, railway applications – specification and demonstration of reliability, availability, maintainability and safety (FDMS);

– automobile: ISO 26262, road vehicles – functional safety;

– electro-medical medical equipment: IEC 60601, general requirements for basic safety and essential performance.

Standard 61508 is a standard on which suppliers rely. In particular, it is possible to acquire a system that guarantees a given SIL. In the following, we explain the principle it proposes for determining an SIS's level of integrity.

The steps in implementing the approach are shown in Figure 8.6.

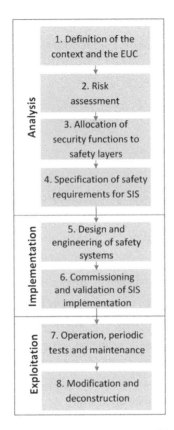

Figure 8.6. *Steps in the process proposed by IEC 61508*

The first step is an initial risk analysis, carried out using one of the methods presented later in this chapter. The objective of this step is to determine the hazards and hazardous events in the system, the sequence of events leading to the hazardous event, the risks associated with the hazardous event, the risk reduction requirements and the safety functions required to reduce the risks. For example, for a boiler that can overheat, the safety function could be to detect the overheating and shut down the burner.

The second step is to define the safety requirements to achieve a tolerable level of risk and translate them into requirements for the security functions in the form of the required SIL. If a LOPA approach (section 8.4.4) is used, functions are assigned to the different layers.

In step 3, the safety requirements of the different SISs, required to implement the safety functions, are defined in terms of SILs.

Step 4 corresponds to the design and actual implementation of the SIS to meet the requirements during commissioning.

8.3. Alignment of safety and security

Cybersecurity is not really taken into account by the IEC 61508 standard. At the time of writing of this standard, ICS cybersecurity was not a major concern. It was considered that the isolation and specific technologies of the control systems were sufficient. The latest version of IEC 61511 (versions 2016 and 2018) (IEC 2016) adds an IT security clause that indicates that the IT vulnerability of SISs must be taken into account, but does not detail this aspect. The IEC 62443 standard, presented in Chapter 7, addresses this problem, and both approaches aim to contain the harmful consequences that can result from a process failure, whether due to a cybersecurity problem or a hazard. It may be interesting to combine the two approaches and a number of methods have been proposed (Kriaa *et al.* 2015; Abdo *et al.* 2018). They will be developed in Chapter 9 with regard to risk analysis.

Indeed, the lifecycle of the two approaches can be aligned (Figure 8.7). The cybersecurity risk assessment (step 5) can be conducted after the functional safety risk assessment (step 2). As explained above, the consequences are similar. However, the likelihood will be assessed with a different metric: a probability for functional security and a qualitative assessment for cybersecurity. This point is further developed in Chapter 9. This risk assessment must also be informed by the description of the SIS specifications (step cyber-2), since these are programmed and may be subject to cyber risks. Conversely, cybersecurity measures must be implemented in the SIS at the security-5 stage. The other steps can be carried out in parallel.

In step 6 of the functional safety implementation, verification is performed on the SIF level of the safety function via the SIL of its components. With regard to cybersecurity, verification is carried out by validating the safety level of each zone.

For the operation and maintenance stage, the actions are similar:

– limitation of access to competent personnel;

– awareness-raising and staff training;

– periodic inspection of safety or cybersecurity measures;

– monitoring of load and failure rates for SISs, while for security, new vulnerabilities are monitored.

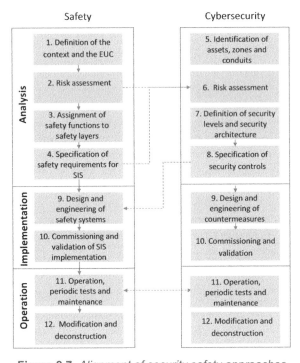

Figure 8.7. *Alignment of security safety approaches*

8.4. Risk analysis methods used in operational safety

These methods of analysis (Flaus 2013) are useful for an integrated safety and cybersecurity approach, as presented in Chapter 9.

8.4.1. *Preliminary hazard analysis*

The objective of the preliminary hazard analysis (PHA) is to identify *a priori* the feared dangerous events, that is, before an accident occurs. This is done by reviewing the elements present in the facility and the activities that take place there to identify sources of danger and to examine the possibility that a feared event may occur.

In the case of a system, its components and operation are examined. If necessary, the harmful consequences are examined. The significance of the various risks identified is generally assessed in a qualitative or semiqualitative manner.

The steps of a PHA are as follows:

1) preparation of the analysis: definition of the context, research of information and observations;

2) description and modeling of the installation;

3) identification of hazards and feared events; a feared event characterizes a mechanism or phenomenon that causes harmful consequences;

4) analysis of dangerous situations and causes that could generate the feared event;

5) analysis of the consequences in terms of damage to potential targets;

6) search for existing barriers;

7) assessment of severity and frequency or likelihood, in a qualitative or semiqualitative manner;

8) proposal of new barriers as needed;

9) drafting of the analysis report.

The results are often presented in a tabular format (Figure 8.8).

System	Causes	Dangerous situation: element and drift	Feared event	Consequences: Target and damages	L	S	RL	Existing Safety Barriers	L'	S'	RL'	Safety Barriers to be added	L''	S''	RL''
	optionnal					optionnal						optionnal			

L: likelihood
S: severity
RL: risk level

Figure 8.8. *Standard PHA table*

8.4.2. *Failure Mode and Effects Analysis*

Failure Mode and Effects Analysis (FMEA) is, like the PHA method, an *a priori* risk analysis method. However, unlike the PHA method, its objective is not to search for feared events that could lead to damage, but to examine all of a system's failure modes, whether they produce damage or only malfunctions, and then to analyze their effects and investigate their causes (Figure 8.9). When, in addition, the criticality of each of these modes is evaluated, then we speak of the FMECA method (Failure Modes, Effects and Criticality Analysis). However, it is often the term FMEA that is used to refer to the method in general terms.

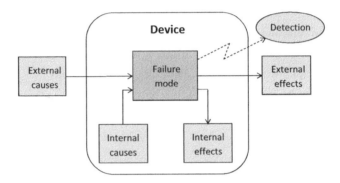

Figure 8.9. *Failure mode*

The effects of a failure mode are of all types and, for the most part, they do not cause damage to a target, but rather an interruption of the system's mission, or an inability to meet the needs for which it was designed. Consequently, the FMEA method is used in many fields other than safety risk analysis, in particular as part of a quality approach (ISO 9000 standard) or for product design.

The steps of an FMEA are as follows:

– preparation of the analysis: definition of the context, research of information and observations;

– description and modeling of the installation;

– analysis of failure modes, their effects and causes;

– probability assessment;

– severity assessment;

– assessment of the analysis and measures to be taken;

– drafting of the analysis report.

The results are provided in the form of a table (Figure 8.10).

System	Entity or function	Failure modes	Possible causes	Possible effects	S	P	Means of detection	Existing preventative meas	Notes

Figure 8.10. *Example of an FMEA table*

8.4.3. HAZOP

The HAZOP (Hazard and Operability Study) method was developed in 1963 by ICI for the chemical industry. The first publications describing this method date back to the 1970s. The principle of this method consists of systematically analyzing all deviations of the operating parameters of the various elements or stages of the operating mode, and then, in searching for those that could potentially lead to the occurrence of a dangerous phenomenon. It is particularly suitable for systems involving material and energy flows. These flows are characterized by parameters such as flow rate, temperature, pressure, level and concentration, whose deviations are examined. By its nature, this method is based on examination of flowsheets or Piping and Instrumentation Diagram.

The study points are analyzed using the following steps:

– first, choose a study point. It generally includes equipment and its connections, and was identified during the modeling stage;

– select an operating parameter;

– retain a keyword and generate a deviation; keywords are taken from a list defined by the HAZOP method, e.g (MORE THAN, LESS THAN, OTHER …);

– examine the consequences of this deviation. If they are not harmful, do not analyze the causes and means of prevention;

– identify the potential causes of this deviation;

– examine the means of detection for this deviation as well as those intended to prevent its occurrence or limit its effects;

– assess the severity and likelihood of this deviation (optional);

– propose, if necessary, recommendations and improvements;

– retain a new keyword for the same parameter, generate a new deviation;

– when all keywords have been considered, analyze a new parameter.

The results are provided in a tabular format (Figure 8.11).

System	Entity of function	Deviation	Causes	Consequences	S	P	Means of detection	Existing preventative measures	Notes
					optional				

Figure 8.11. *HAZOP table*

8.4.4. *Layer Of Protection Analysis*

The Layer Of Protection Analysis (LOPA) method was developed in the late 1990s by the Center for Chemical Process Safety. The objective of this method is to evaluate the level of risk control with existing barriers on a system and to decide if additional barriers should be added. It is defined by its authors as a simplified semiquantitative method, because it is based on levels defined by powers of 10 to represent probabilities, frequencies and gravity levels.

A key concept in the LOPA approach is the concept of an Independent Protection Layer (IPL, Figure 8.12). The method aims to determine whether sufficient levels of protection have been put in place to ensure a tolerable level of risk.

An IPL is defined as a system that performs a safety function, either actively or passively, and that must:

– detect and prevent or mitigate the consequences of specified feared events, such as loss of containment or runaway reaction;

– be independent of all other layers of protection associated with the feared event;

– be reliable in order to reduce the risk by a specified level;

– be verifiable in such a way as to allow periodic validation of the safety functions it provides.

In concrete terms, an IPL may be a standard operating procedure, the process control system, an alarm with operator acknowledgment, an instrumented safety system, a valve, a fire detector, a sprinkler, etc.

The LOPA method is intended to be a method capable of ensuring that the level of risk is controlled to an acceptable level (Figure 8.13) by adding measures that must be successively overcome. It is similar to the defense-in-depth approach described in Chapter 3.

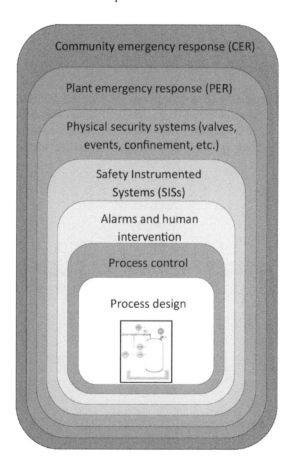

Figure 8.12. *Independent Protection Levels (IPL). For a color version of this figure, see www.iste.co.uk/flaus/cybersecurity.zip*

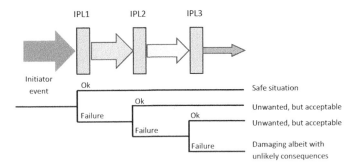

Figure 8.13. *IPL and risk management. For a color version of this figure, see www.iste.co.uk/flaus/cybersecurity.zip*

8.4.5. *Fault trees and bowtie diagrams*

The fault tree is a model that highlights the logical combinations of faults that can lead to the main event of interest, called the "top event". Its graphic representation is tree shaped. The various events are connected by logical connectors. This representation is very similar to that of attack trees (Chapter 9).

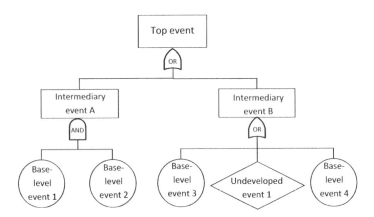

Figure 8.14. *Example of a fault tree*

In operational safety, the fault tree is used for quantitative analysis. A probability is associated with each base event and, from the structure of the fault tree, it is possible to calculate the probability of the top event.

Another graphical representation very often encountered is the bowtie diagram. The idea is to represent for each feared event all the causes and consequences (Figures 8.15 and 8.16). Causes and consequences can have several levels. Barriers are also represented on this type of diagram, which makes it possible to visualize the level of risk control for the scenario in question.

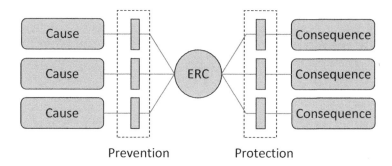

Figure 8.15. Bowtie diagram

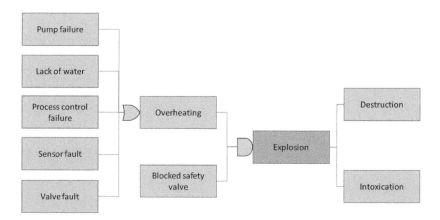

Figure 8.16. Bowtie diagram

9

Risk Assessment Methods

9.1. Introduction

Managing cybersecurity in an industrial control system (ICS) involves an important proactive phase, which consists of implementing measures to reduce the likelihood of an attacker generating damage to the system. This approach is described by the risk management process of ISO 27005. In order to be able to concentrate efforts and resources on important aspects, the first step is to carry out an analysis of the risks that the installation or system runs, and those that it causes to its environment.

When considering the cybersecurity of an industrial installation, one must consider two types of risk analysis: that of the control–command system or that of the physical system. The first is a risk analysis of the industrial information system, which will be carried out using an analysis method developed for information systems; the second is a so-called "industrial" or operational safety risk analysis, for which a method such as that described in Chapter 8, such as a Preliminary Hazard Analysis (PHA) or a Hazard and Operability Study (HAZOP), will be used. These two analyses are of course not independent, since the consequences of an attack on the ICS can trigger a dangerous scenario in the physical part.

Several approaches are then possible to assess risk for the overall system:

– carry out risk analysis for the industrial information system and consider loss of control over the installation as an impact, the severity of this loss of control being assessed approximately;

– carry out risk analysis for the physical process and take into account the causes of cyber-attacks on the ICS, the likelihood of these causes being roughly characterized;

– carry out both analyses and combine them to more accurately assess the overall risk generated by the ICS and installation set;

– carry out the risk analysis of the physical process, identify the causes due to cybersecurity and provide "non-attackable" barriers to ensure the safety of the whole.

The choice of approach depends on the issues and context. The remainder of this chapter presents information system risk analysis methods and their adaptation to the case of industrial systems, followed by adaptation of the methods presented in Chapter 8 to take into account cybersecurity aspects and the approaches combining them.

9.2. General principle of a risk analysis

9.2.1. *General information*

A number of risk analysis methods have been developed for IT systems. Table 9.1 presents the most commonly used. Most of the methods follow the steps proposed in ISO 27005 and therefore have some common points. This general approach is presented in Figure 9.1. It is a top-down approach. Carrying out a risk analysis is a difficult task: it is necessary for it to be precise while maintaining an overview covering all risks and situations, while remaining homogeneous. To make the process as rigorous as possible, the contribution of a method consists first of all of proposing a systematic approach composed of fairly simple steps, which generally consist of inventorying the system, representing it in one way or another, then identifying the risks based on generic lists and evaluating them using a well-defined scale. The purpose of generic lists is to ensure that the analyst does not forget important points. Depending on the methods, they can describe component categories, threats, vulnerabilities, possible impacts or even generic scenarios.

These methods can be applied at different levels: at the global level, to conduct an impact analysis, at the intermediate level, by examining the company's organizational processes, or at the technical level.

Name	Designer	References
EBIOS	ANSSI, 1995, 2010, 2018	https://www.ssi.gouv.fr/uploads/2018/10/guide-methode-ebios-risk-manager.pdf
MEHARI	CLUSIF, 1995	http://www.meharipedia.org/home
OCTAVE	Carnegie Mellon, 1999	http://www.cert.org/octave/osig.html
CORAS	UiO/SINTEF, 2001	http://coras.sourceforge.net/
MAGERIT	Spanish Ministry for Public Administrations, 1997	http://www.csi.map.es/csi/pg5m20.htm
SP800-30	NIST, 2002	http://www.csrc.nist.gov
CRAMM	British CCTA, 1985	http://www.cramm.com
TARA	MITRE, 2011	https://www.mitre.org/publications/technical-papers/threat-assessment–remediation-analysis-tara

Table 9.1. *Main methods of information security risk analysis*

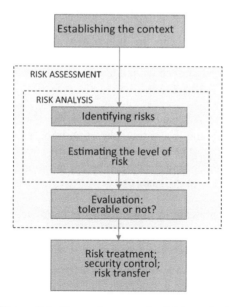

Figure 9.1. *General approach of risk analysis*

At the global level, impact analysis focuses on the different categories of danger: fire, natural phenomena, terrorism, workplace violence, pandemic, cybersecurity, various breakdowns, supplier defection, etc. It examines their impact at the company level: material damage, operating losses, loss of customers, financial losses, environmental contamination, loss of trust in the organization, fines and penalties, lawsuits, etc. With regard to cybersecurity, this step can be a prerequisite for highlighting the importance of IT security risks in order to involve management in a cybersecurity control approach.

To implement a risk management approach, the analysis is carried out at the process level in order to ensure compliance with good practices or international standards for information security management.

A more detailed analysis can also be carried out at the technical level, the usual level of analyses for operational safety. For example, threat modeling can be carried out at this level with a model such as STRIDE[1], the objective of the analysis being to assess vulnerability, taking into account technical measures.

The analysis can also be mixed, as proposed in the NIST SP 800-30 guide or IEC 62443. The level of risk is assessed in relation to technical measures, for example, the presence of an antivirus, and also in relation to the organizational process that manages and updates the deployment.

It is important to keep in mind that "information security risk management, like risk management in general, is not an exact science. It brings together the best collective judgments of individuals and groups in the organization" (NIST SP 800-39).

The objective of risk analysis methods is therefore only to provide an approach to identify and assess the risks of an information system that is as systematic as possible. The action of implementing a method, which makes it possible to better understand the system and to question its weaknesses, is as important as the result of the analysis.

The following presents in more detail the main steps of a risk analysis method.

---

1 STRIDE is a threat classification model developed by Microsoft to address IT security threats. It provides a mnemonic for security threats in six categories: spoofing, tampering, repudiation, disclosure, denial of service and elevation of privilege.

9.2.2. *Setting the context*

This phase consists of defining the objectives of the study, the perimeter of the installation studied and the important assumptions. It corresponds to the part of the definition of the context of the risk management process defined in ISO27005 (Chapter 3), and generally includes the steps of inventorying system elements, defining metrics (how to measure the level of risk) and the threats taken into account.

A preliminary step is to carry out a system inventory, that is, to identify all the hardware and software components used for information processing. These components are referred to as "assets" in the following. For an ICS, this set is delimited by the connection to the IT network or Internet on one side, and by the connections to the physical process via sensors and actuators on the other.

This inventory includes workstations and servers, specific equipment (PLCs, HMI, etc.), but also means of information transfer (network, router, firewall, etc.). The network may be wireless, and some equipment may be mobile, especially if the company has a BYOD (Bring Your Own Device) policy. For each system, the inventory should provide the type of operating system with its version number and a list installed applications with their version numbers. Such an inventory can also be very useful to identify users and their privileges.

The work to be done at this stage can be considerable, even though some tools exist (section 11.1). Together, they make it possible to draw up a mapping of the installation in the form of a diagram (Figures 10.1 and 10.2) or in the form of a table listing the assets.

In the context of an ICS, it is also important to identify the links between the control system and the controlled system. To do this, it is useful to carry out a functional analysis of the physical system and to identify the support functions required by the control system. It is also necessary to identify the safety functions.

For example, in a chemical production system, we try to identify the main functions, which can be, for example, filling, emptying and temperature regulation. A safety function may be to limit the temperature. Then the devices performing these functions are identified. The final step is to find how they fit in the mapping and how they interact with other information system assets.

Assets are affected by different threats and are used to carry out business processes. In the case of ICS, the production and safety functions are addressed and the assets used to perform these functions must be identified. This allows impacts to be determined systematically.

The metrics (section 3.3.7) that measure likelihood, severity and risk level are defined in this step.

9.2.3. Risk identification

The inventory of assets and services makes it possible to establish a list of the company's critical assets and processes. The impact analysis is carried out by examining the consequences on each of the security objectives: availability, integrity and confidentiality. For control systems, which act on real systems, the consequences of a loss of control of the physical system must be taken into account.

In addition to direct impacts, there may also be indirect impacts, which often result in financial losses: losses in production quantity and quality due to untimely shutdowns, image degradation due to poor service or disclosure of confidential data.

When the impact is identified, the question is whether the associated feared event can occur. To do this, we seek to identify the causes that could lead to it, i.e. to identify the potential vulnerabilities or weaknesses of the various elements of the installation involved in implementing the potentially impacted process or element, taking into account the sources of threats considered. This identification provides a list of scenarios.

For example, for the feared "loss of equipment availability" event, the causes may be a denial-of-service (DoS)[2] attack, a power failure or a flooding of the premises.

Each of these scenarios can be associated with a source of threat: for example, the DoS attack may come from a generalized attack or from an attacker specifically targeting the installation.

---

2 See Chapter 4 for more details on this type of attack.

9.2.4. *Estimation of the level of risk*

The level of risk is assessed based on the likelihood and severity of the impacts using a risk matrix. Severity is assessed by considering the importance of the impact with respect to the defined metric. The likelihood of each scenario can be evaluated, either on the basis of the event that initiated the feared event, or on the basis of an analysis of the vulnerability and level of the attacker.

Different scales can be used: some have been presented in Chapter 3. Others metrics are given in section 9.3, which presents the expression of needs and identification of security objectives (EBIOS) or section 9.5.2, which describes the cyber PHA approach.

9.2.5. *Risk assessment and treatment*

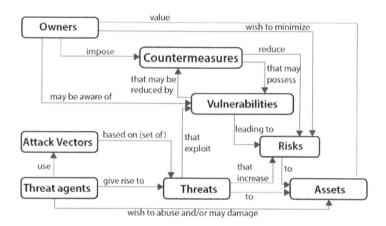

Figure 9.2. *Controls and risk*

At this stage, the list of risks has been drawn up and each of them has been evaluated. The next step is to decide which actions to take or not to take depending on the level of risk.

Some analysis methods propose lists of generic security measures. The objective is of course to reduce vulnerabilities (Figure 9.2). The different lists of measures that can be used are as follows:

– ISO 27002 (ISO/IEC 2013), which includes 14 categories of 35 objectives and 114 risk reduction measures;

– the NIST guide SP800-53 (NIST 2014), a catalogue of 163 measures, classified according to whether they concern the identification, protection, detection, response and recovery phases, and which can be grouped according to whether they concern management, technical or operational aspects;

– IEC 62443-2-5 (ISA 2018) dedicated to ICS;

– the NIST guide SP 800-82 (Stouffer *et al.* 2015) dedicated to ICS (Appendix 3);

– the ANSSI guide, detailed measures (ANSSI 2013a), dedicated to ICS (Appendix 4).

There is also the OSA repository, based on the NIST repository[3], or OWASP for applications[4].

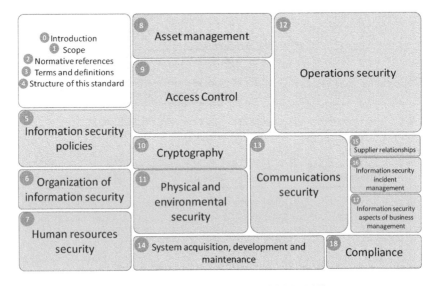

Figure 9.3. *Security controls (ISO27002)*

---

3 Available at: http://www.opensecurityarchitecture.org/cms/library/0802control-catalogue.
4 Available at: http://www.opensecurityarchitecture.org/cms/library/0802control-catalogue.

9.2.6. *Tailor-made approach and ICS*

From the elements presented above, it is possible to build a tailor-made method adapted to a company or a problem. It is sufficient to repeat the steps presented and adapt them, essentially specifying metrics and guide lists of threats, vulnerabilities and control measures, based on the generic lists mentioned above.

In addition, when it comes to ICS or cyber-physical systems, the most important security requirements are availability, so as to guarantee efficient real-time control and guarantee the proper functioning of safety functions, as well as the integrity of data, whether in databases stored on disk, exchanged between equipment, or stored in the working memories of PLCs and various devices. These data consist of the control programs, associated variables and configuration.

9.3. EBIOS method

One of the most widely used methods in France, recommended by ANSSI (ANSSI 2018) is the EBIOS[5] method that we will describe in more detail.

The general approach of this method is divided into five steps (Figure 9.4). It broadly follows the steps of the ISO 27005 standard. The latest version of this method, called EBIOS RM (EBIOS Risk Manager), aims to simplify the implementation of the approach. Unlike the previous version, it makes it possible to implement a preliminary analysis step that may be sufficient to define the priority axes of the risk control approach. It does not require a detailed assessment of the scenarios and their likelihood. The detailed analysis will only be carried out in the case of a complete and detailed risk study.

In addition, this new version aims to create a security management methodology combining risk analysis with compliance to a checklist provided by a standard. The compliance approach is used in the initial step (workshop 1) to determine the potential security gaps.

---

5 In French, *Expression des Besoins et Identification des Objectifs de Sécurité* (expression of needs and identification of security objectives).

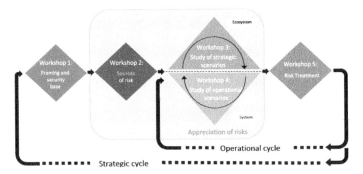

Figure 9.4. *Steps of the EBIOS method. For a color version of this figure, see www.iste.co.uk/flaus/cybersecurity.zip*

9.3.1. Workshop 1: framing and security base

This workshop is divided into four stages:

a) defining the framework for the study;

b) defining the business and technical scope;

c) identifying feared events and their severity level;

d) determining the security base: list of applicable standards, state of application, identification and justification of discrepancies.

9.3.1.1. Definition of the study framework

This step consists of specifying the objectives of the study, which may be, for example, the implementation of a risk management process related to information security, the approval of an information system or the completion of a risk level inventory. The level of granularity and the choice of workshops to be implemented for the application of the EBIOS method depend on these objectives.

If the objective is compliance analysis, workshops 1 and 5 are used. If the objective is to carry out a preliminary risk analysis, the study will be carried out by selecting workshops 1 (macroscopically), 2, 3 and 5 (step a). If the objective is to carry out a detailed study, all workshops are implemented.

After defining the objectives, it is then necessary to identify the participants in the different workshops and their roles. Finally, a temporal planning of the different stages is carried out.

9.3.1.2. Definition of the business and technical scope

The perimeter is defined by listing:

– the missions, or in other words, the purpose(s) of the installation (for example, to manufacture a given product);

– business values, also called essential (or core) assets, which are the processes, functions and sets of information necessary to carry out the missions (for example, managing a production chain). For each business value, security requirements are identified (AIC, section 4.1.1). A scale such as the one proposed in Table 9.3 can be used to characterize these requirements;

– the necessary support assets for business values. These are the concrete elements that make it possible to carry out the process or manage the information. They belong to different categories, including:

    - equipment: computers, communication delays, USB sticks, hard disks, etc.;

    - software: operating systems, email, databases, business applications, etc.;

    - computer communication channels: cables, Wi-Fi network, optical fiber, etc.;

    - people: users, IT administrators, decision makers, etc.

The set can be grouped into a table (Table 9.2).

Missions	Business values (core assets)	Security requirements (AIC)	Supporting properties (the most important) and responsible

Table 9.2. *System description*

Scale levels of availability	Detailed description of the scale
1 – Not very important	The availability of the property is not very important
2 – Important	The property may be unavailable for a limited time
3 – Critical	The proper functioning of the asset is critical
Levels of the integrity scale	**Detailed description of the scale**
1 – Detectable	May have a loss of integrity if alteration is detected
2 – Controlled	May not be intact, if the alteration is identified and the integrity of the essential property is found
3 – Fully integrated	Must have strict integrity
Levels of the confidentiality scale	**Detailed description of the scale**
1 – Public	Available to everyone
2 – Limited access	Available only to staff
3 – Confidential	Available only to identified persons

Table 9.3. *AIC scales*

9.3.1.3. *Identification of feared events and their severity level*

The feared events are obtained by considering each criterion for each business value: they represent deviations from security requirements. For example, for a property that must be confidential, a feared event is the loss of confidentiality (access by unauthorized persons).

Levels of the seriousness scale	Detailed description of the scale
1 – Negligible	No impact
2 – Limited	Reversible impact
3 – Significant	Problematic but manageable impact
4 – Catastrophic	Significant, irreversible impact

Table 9.4. *Severity scale*

In a second step, for each feared event, the possible impacts are identified for each type of impact. The types of impacts are as follows: impacts on operations, human impacts, impacts on properties, environmental impacts, etc. The severity of these impacts is assessed on the scale in Table 9.4 and the highest value is used.

The whole is summarized in a tabular format (Table 9.5).

Missions/ business value	Feared event	Impacts	Severity

Table 9.5. *Study of feared events*

9.3.1.4. *Determination of the security base*

To determine the security base, it is necessary to identify the list of measures that are supposed to apply. These come either from a normative standard or guide with which the system must comply, or from a list of security controls established during a previous risk analysis.

The applicable standards include the different standards (ISO 27000, rules of computer hygiene, etc.).

If the system under analysis exists, the status of the measures' implementation should be assessed and can be summarized in a table (Table 9.6). This is done by identifying existing security measures. For each identified media asset, it is necessary to identify the existing security measures (logical or physical security product, configuration, rules, procedures, etc.) and to evaluate deviations from the baseline.

Each security measure can usefully be categorized according to the line of defense (preventive, protective or recovery) to which it belongs. This will later facilitate, if necessary, setting of additional security measures according to the principle of defense in depth.

Type of reference	Name of reference	State of application	Deviations	Justification of deviations
Rules of good practice	Guide or standard reference	Applied with restrictions		

Table 9.6. *Security base*

At the end of workshop 1, we get an overview of the system, issues and deviations from the baseline.

9.3.2. *Workshop 2: sources of risk*

This workshop aims to identify the sources of risk (SR) and the target objectives (TOs) for these sources. The source/objective pairs are then evaluated to determine their level of relevance. For example, a source may be a competitor whose objective is to steal information or, in the case of an ICS, a cyber-terrorist who seeks to create an industrial disaster by making the installation uncontrollable.

The study is based on the missions and business values obtained at the end of Workshop 1 and it consists of looking for sources of risk that could affect the organization's missions, and what the overall objective of the risk source may be.

The steps of this workshop are as follows:

1) identification of risk sources and objectives: a list of generic sources (Tables 4.1 and 4.2) can be used and the sources selected can be contextualized. For example, an external source with significant resources may be a competitive or cyber-terrorist source;

2) assessment of SR/TO pairs: each SR/TO pair obtained in the previous step is assessed in order to select those that are most relevant. The criteria usually used are the motivation of the source, its resources in financial and technical terms, its level of skills, as well as its activity within the scope of the study or the field concerned. The whole can be summarized in the form of a table;

3) selection of pairs used in the rest of the analysis: based on the information from the previous step, SR/TO pairs are evaluated in terms of relevance. The most relevant couples are selected and care is taken to make them sufficiently distinct from each other.

Sources of risk	Target objectives	Motivation	Resources	Activity	Relevance
Competition	Theft of information	+++	++	+++	High
Cyber-terrorists	Industrial accident	+	+++	+	Low

Table 9.7. *Risk sources and target objectives*

9.3.3. *Workshop 3: study of strategic scenarios*

The objective of workshop 3 is to determine the attack paths that sources of risk can take at the ecosystem level, i.e. at the level of stakeholders gravitating around the subject of the study. These stakeholders, consisting of partners, various service providers, suppliers and customers, may be particularly vulnerable and more and more attack operating modes are exploiting these weaknesses.

The attack scenarios established during this analysis are called strategic scenarios. This workshop can be seen as a preliminary risk analysis to determine:

1) an overview of the digital threat to the ecosystem and critical stakeholders;

2) the associated strategic scenarios and feared events;

3) the security measures chosen for the ecosystem.

9.3.3.1. *Construction of the mapping of the digital threat of the ecosystem and selection of critical stakeholders*

A stakeholder is considered critical (CCS) when they are likely to be a relevant vector of attack, for example, because of its privileged digital access to the object under study, its vulnerability or its exposure.

The work to be carried out in this step consists of establishing the list of stakeholders (customers, partners, service providers, suppliers, etc.) based on the mapping of the information system and its ecosystem. For each stakeholder, it is necessary to analyze the level of threat on the subject of the study. Stakeholders are assessed on the basis of exposure criteria (dependence, penetration) and on the level of risk control related to cybersecurity (maturity, trust).

9.3.3.2. *Development of strategic scenarios*

The previous step was used to build the threat mapping and identify critical stakeholders. The objective is now to build the scenarios that will allow an attacker to achieve his objectives. These scenarios, called strategic scenarios, are identified by using the list of SR/TO pairs and searching for each of them:

– the business value(s) of the organization that the source must aim to achieve its objective;

– critical ecosystem stakeholders with privileged access to the business values that are relevant to attack.

Then, for each business value and critical stakeholder pair, the steps to achieve the feared event are determined. This sequence of events refers to description of the scenario. It can be represented as a tree or an attack graph (section 9.4).

The severity level of a scenario is obtained by taking the severity level of the feared event generated. As part of this scenario development phase, the list of feared events to be considered is the one established in workshop 1 and can be modified and updated as necessary.

9.3.3.3. *Definition of ecosystem security measures*

During this step, the objective is to define the security measures that will reduce the possibility of carrying out the identified scenarios. These measures may concern threats induced by critical stakeholders (e.g., reducing dependence on a subcontractor) or influencing the development of the strategic scenario.

9.3.4. *Workshop 4: study of operational scenarios*

The objective of workshop 4 is to determine in a more concrete way the operational attack paths that the source must use to implement the strategic scenarios identified in workshop 3. The approach focuses on the support assets associated with the various missions and business values.

This workshop is divided into two stages:

1) development of operational scenarios: for each strategic scenario, an operational scenario is constructed. Scenarios can be represented by attack trees (section 9.4);

2) likelihood assessment: the strategic scenario is a schematic diagram for conducting the attack, whereas the operational scenario specifies in detail the path followed: its likelihood can therefore be assessed. One approach can be to assess the elementary likelihood of each action by taking into account the measures identified by the security base, and then combine them to obtain the overall likelihood. It is also possible to perform a direct assessment of likelihood.

During this workshop, it may appear that some strategic scenarios need to be modified, or even added or deleted.

Levels of the likelihood scale	Detailed description of the scale
L4 – Almost certain	The source of risk will certainly achieve its intended objective according to one of the paths of the scenario. The likelihood of the scenario is very high.
L3 – Very likely	The source of risk can probably achieve its intended objective according to one of the paths of the scenario. The likelihood of the scenario is high.
L2 – Likely	The source of risk is likely to achieve its intended objective according to one of paths of the scenario. The likelihood of the scenario is significant.
L1 – Unlikely	The source of the risk is unlikely to achieve its intended objective according to one of paths of the scenario. The likelihood of the scenario is low.

Table 9.8. *Likelihood scales*

9.3.5. *Workshop 5: risk treatment*

The purpose of this workshop is to synthesize the identified risk scenarios and define security measures and a risk control plan.

The steps of this workshop are as follows:

1) perfroming of the synthesis of all risk scenarios;

2) definition of the risk treatment strategy and of the security measures;

3) assessment and documentation of residual risks;

4) implementation of the risk monitoring framework.

9.3.5.1. *Performing the synthesis of all risk scenarios*

Risk assessment is carried out by examining which scenarios generate the different feared events. For each combination of a feared event and a scenario that generates it, a risk is obtained, characterized by the severity of the impact of the feared event and the likelihood of the scenario. The scenarios generating a risk are those concerning a support asset on which the business value (essential asset) depends.

It should be recalled that the assessment of likelihood and severity takes into account the existing security measures identified. Those measures make it possible to:

– protect business values or essential assets by preventing the occurrence of the feared event, for example, by reducing the vulnerabilities of the identified support assets;

– reduce impacts by limiting the direct effects of the feared event;

– reduce impacts by increasing the recovery capacity.

The first estimate, called "gross", is made without taking into account existing security measures. The second estimate, so-called "net", is made for the severity and likelihood of each risk, taking into account the effect of existing security measures, if any.

The different risks described by a couple (scenario, feared event) can be positioned on a risk matrix (Figure 9.5), which makes it possible to define their levels.

Risk Assessment Methods 231

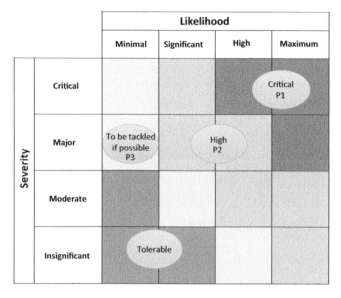

Figure 9.5. *Risk matrix. For a color version of this figure, see www.iste.co.uk/flaus/cybersecurity.zip*

It may be useful to check the coverage of feared events, i.e. to check that at least one scenario has been constructed for each feared event.

9.3.5.2. *Definition of the risk treatment strategy and of the security measures*

Depending on the level of risk obtained, a treatment option should be chosen and the security measures to be implemented have to be defined. For the risk matrix presented in Figure 9.5, the strategy can be as follows:

– level "P4: low" (green): the risk is accepted and no action is taken;

– level "P3: moderate" (yellow): the risk is monitored via indicators and actions are implemented as fast as possible;

– level "P2: high" (orange): the risk is monitored via indicators and actions are implemented quickly;

– level "P1: critical" (red): the activity is stopped and security measures are implemented immediately.

For levels P1, P2 and to some extent P3, security measures are defined to reduce the risk. They complement the ecosystem actions taken in workshop 3. The proposed measures may reinforce those put in place under the security base or may be designed specifically for the scenario under analysis. They aim to reduce its likelihood but can also affect the severity of the feared event.

All measures are documented and for each of them, we identify:

– its category (preventive, protective or recovery);

– its possible reference in relation to the standards used;

– the support asset that carries it in order to facilitate the optimization of security measures, its owner being *a priori* responsible for application of the security measure;

– the person responsible for implementation;

– its cost;

– the deadline for the implementation.

The whole process makes it possible to develop the continuous security improvement plan (CSIP), which is staggered over time and structured. In a general manner, each action in the plan should verify the SMART criteria:

– S: specific (one actor, one domain at a time);

– M: measurable (definition of the indicator of progress);

– A: achievable (possibly in several steps, with the necessary resources);

– R: realistic (depending on the actors, their capacities);

– T: time related (with a deadline, within a defined period).

Security control	Related risks	Responsible	Obstacles and difficulties for implementation	Cost/ complexity	Completion date
Operator awareness	R_i	IT security			3 months

Table 9.9. *Plan of measures to be implemented*

9.3.5.3. *Residual risk assessment and documentation*

This step consists of carrying out a residual risk analysis. This involves identifying and estimating the residual risks that will remain, once each security measure has been implemented, and verifying that they are ready to be accepted with full knowledge of the facts. It is also useful to check that the security measures do not generate new risks.

Residual risks are documented by describing the scenario, the associated feared event and security measures to reduce the initial risk. A new assessment of likelihood and severity must be performed.

9.3.5.4. *Implementation of a risk monitoring framework*

This step consists of defining a monitoring framework based on key risk indicators (KRI). These indicators make it possible to verify the effectiveness of the measures taken and adaptation to the state of the threat.

A continuous security improvement process can be set up (steering committee, definition of a risk analysis update cycle, etc.).

9.3.6. *Implementation for ICS*

The EBIOS method was originally developed for information processing systems and not for control command systems. Essential or Core assets usually considered were therefore essentially data that could be inaccessible, corrupted or disclosed, or information processing services that could be unavailable.

In the context of an ICS, business values or essential assets are mainly functions that make it possible to control the physical system and associated data: these are the control and regulation functions provided by the basic process control system (BPCS), as well as the safety functions provided by the safety instrumented system (SIS).

These functions access the state of the physical system via information from the sensors, process it, store part of it and then send commands back to the physical system via the actuators. The important information consists of the value of the measured values, the stored values, the programs of the PLCs that perform the processing and the various configuration information. All this information and the associated functions must therefore have a perfect integrity.

In addition, the system must respond in real time, so it must remain available, with a very short maximum unavailability time, in the order of milliseconds, since the physical system is constantly evolving and the safety functions must react almost instantly.

As far as confidentiality is concerned, the stakes are lower, even though there may be a requirement for confidentiality on production information, and though PLC programs and, in general, documents describing the installation should not be made available to malicious persons who could use them to detect vulnerabilities.

In addition, from a practical point of view, it may be interesting to use a simplified table (Figure 9.6).

Core assets	Feared event	Security requirement	Impacts	Threat sources	L (likelihood)	S (severity)
Supervision	Tampering of measured data	3. Integrity	Supervision screens out of step with real values Uncontrolled production	Internal human source, with malicious intent and limited capabilities External human source, with malicious intent and limited capabilities	3. High	3. Significant

Figure 9.6. *Example of an EBIOS analysis table. For a color version of this figure, see www.iste.co.uk/flaus/cybersecurity.zip*

9.4. Attack trees

In the methods mentioned above (Table 9.1), the scenarios are described in the text form. To carry out a detailed analysis, it may be useful to use a more formalized approach, which will better highlight attack paths and security measures. One solution is to use a model inspired by the fault tree and called the "attack tree".

Attack trees (Figure 9.7) are a graphical representation of combinations and sequences of events that describe the different ways in which a system can be attacked. The top node of the tree represents the ultimate goal of the attack. This is described as a combination of subobjectives, each of which may represent an intermediate state of the installation, a system vulnerability or an attacker's action. These different facts are linked together by logical

AND and/or OR connectors. It may also be possible to show the security measures on the attack tree.

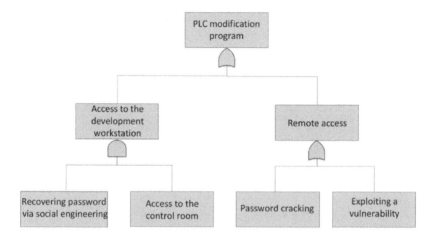

Figure 9.7. *Example of an attack tree*

This representation is similar to that of the fault tree presented in Chapter 8, first adapted in the field of information in 1999 (Schneier 1999) and then formalized (Mauw and Oostdijk 2006). A first application for ICS vulnerability analysis was developed in 2004 (Byres *et al.* 2004).

The graphical and structured tree notation provides an interesting overview and is also suitable for systematic threat analysis tools. It is also a useful tool for analyzing security measures.

Attack trees can be generalized in different ways. It is possible to consider an event graph, called an "attack graph", capable of representing multiple scenarios and generating several attack trees (McQueen *et al.* 2005).

It is also possible to add defense mechanisms in the tree, which makes it possible to reveal possible paths for an attack and existing protection (Kordy *et al.* 2011). A measure is represented by a node, in the form of a rectangle, and is connected on a link between two nodes meaning that the measure reduces acts on this relation. It is also possible to show the elements constituting this measure (Figure 9.8). In the same way that attack trees are similar to fault trees, these attack–defense trees are very similar to bowtie diagrams (Chapter 8). The similarity between the tools allows a fairly natural convergence of information security and operational safety risk analyses.

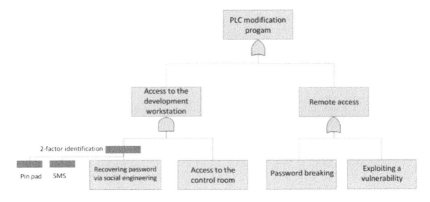

Figure 9.8. *Example of an attack–defense tree*

9.5. Cyber PHA and cyber HAZOP

9.5.1. *Principle*

Most often, as explained in Chapter 8, industrial systems are subject to risk analysis. The most commonly used method is the PHA method, but the HAZOP, FMEA or sometimes other methods are also used. This analysis is carried out as part of a functional safety study or for a hazard study.

In the traditional approach, cybersecurity issues are usually not taken into account. The causes of the hazardous situations considered are component failures, human errors and abnormal events from the environment.

To take into account the risks associated with cybersecurity in these studies, it is necessary to consider new categories of causes for the deviations or dangerous situations identified. In the context of a PHA, for each feared event, in addition to causes related to technical failures or human errors, deliberate unintentional operation caused by computer systems will be considered. In parallel, for each safety instrumented barrier, i.e. one based on a computer system or an embedded system, malfunctions resulting from cybersecurity will be considered. Analysis of industrial risks or operational safety therefore makes it possible to express specific requirements in terms of IT security by highlighting the IT elements that can be a source of malfunctions on the physical system.

In general, wherever one of the causes may be related to a problem of process control (sensors, actuators, computation of command for actuators), there will be a requirement for integrity of the used data. In addition, if the system requires continuous time control and monitoring, there will be a requirement for availability. If the system is an SIS, then there will be an even greater requirement for availability.

The overall risk level of the facility can then be assessed by taking into account security and cybersecurity and using a cyber fault tree or cyber-bowtie approach (Abdo et al. 2018) (section 9.6).

To perform a cyber PHA, a double rating must be used for likelihood. Indeed, the likelihood of computer attacks and the likelihood of failures are not of the same order of magnitude. In the first case, it is rather high, and in the second case, it represents a probability of failure and is therefore relatively low. Combining the two magnitudes may hide one of the two components, so it is better to use a double scale.

A first rating scale is used for causes of physical origin, whether they are initiating events, component failures, human errors or barrier failures. It corresponds to the likelihood usually used in a PHA for operational safety. It is noted P in the following, and it can be qualitative, for example with five levels that we will identify with a letter, ranging from A, very likely, to E, almost impossible. It can also be represented by a probability, real number between 0 and 1.

The second rating is used to assess the likelihood of computer attacks, whether they are attacks to perform actions harmful to the system or attacks to put instrumented or computerized barriers (SIS) into a failed state. To distinguish it more easily from the likelihood scale of physical risks, it is noted L and scaled using integers, for example from 1 to 4, as shown in Table 9.8.

A scenario can be generated by physical failures alone, cyber-attacks alone or a combination of both. The idea is therefore to distinguish the following situations:

– those caused by technical or human failure such as in operational safety;

– those caused by computer attacks and therefore related to cybersecurity;

– those with causes from both origins, such as a situation that may occur in the case of an attack, if a physical barrier is deficient or, conversely, a situation caused by a technical failure with an instrumented barrier which is ineffective because of a cyber-attack.

In the first case, the likelihood is assessed by a level on the P scale, in the second case it is a level on the L scale. In the last case, it is a L|P or P|L pairing, which must be considered in a multiplicative way, since a conjunction of the two causes is required. However, since the initiating events are of a different nature, and in order not to mask cybersecurity issues, multiplication is not actually achieved but dual scoring is retained to assess the level of risk via an extended risk matrix (Figure 9.9).

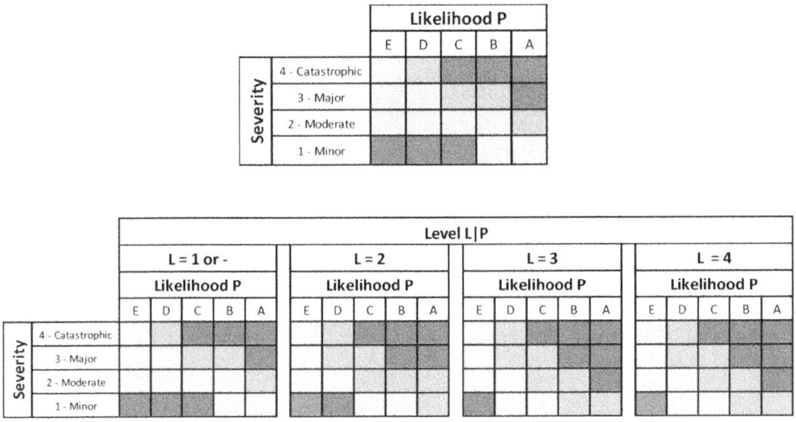

Figure 9.9. *Classic risk matrix (top) and scope. For a color version of this figure, see www.iste.co.uk/flaus/cybersecurity.zip*

This rating system allows us, for example, to be demanding where catastrophic impacts are concerned and, in this case, to base overall security on physical barriers, but also to allow more flexibility for low impacts and, in this case, to base security on both physical and computer-based barriers.

Assessment of the severity of a scenario is carried out by examining the significance of the consequences, in a way similar to what is usually done.

For assessment of the level of risk, in the rating proposed below, trust in the IT part depends on the importance of the consequences. If the impact is maximum, everything is based on P; on the other hand, when the impact decreases, we can release the constraint on the physical barrier and give more trust to the IT part.

This approach can be used to design the risk matrix (Figure 9.9), which has been implemented as follows:

– if there are no cyber causes or barriers, the matrix is identical to that of physical systems risk analyses;

– if the impact is catastrophic, everything is based on the physical systems, so we remain on the same matrix (whatever the value of L, the rating for S = 4 does not change);

– if the impact is major, a cause or barrier based on a computer system is considered to reduce the risk, if its level of likelihood is minimal, because it provides a benefit in terms of controlling the overall risk;

– if the impact is moderate, a significant level of likelihood for cyber risks is considered to decrease the overall level of risk;

– if the impact is minimal, IT-based elements are considered to significantly reduce the level of risk, even if they are vulnerable (for low-impact risks, flexibility is preferred).

9.5.2. Cyber PHA

A cyber PHA can be seen as an extension of a PHA by taking into account cybersecurity issues in the case of a feared event and for barrier failures.

The general approach of a cyber PHA is as follows:

1) preparation of the analysis: definition of the context and scope, gather information and observations; definition of metrics for the likelihood and severity of the impacts considered;

2) description or modeling of the installation, which is broken down into systems. Each system can be seen as a set of components to perform functions (Figure 9.11). Each system receives material and information flows and generates them: it is the system's inputs and outputs that can be identified. It is important to show the information processing systems or functions (BCPS or SIS) and the interfaces with the physical system (sensors, actuators). The more detailed the model, the more systematic the risk analysis will be;

3) analysis of the feared events and their consequences for each system. The feared events correspond to a loss of control of the physical installation;

4) for each feared event, its causes are analyzed. Causes include incorrect control commands or incorrect measurements. These events are called cyber-initiating events, denoted by Cyber-IE;

5) each feared event, and its consequences are analyzed. They can be of different types: human impact, damage to equipment, environmental damage, financial loss. The list of impact categories taken into account is defined in the preparation stage of the analysis. At this stage, the combination of feared event, cause and effect, represented on a line of the PHA table (Figure 9.13), can be seen as a scenario (Figure 9.10);

6) analysis of existing barriers, determining if they are cyber-attackable, i.e. if they work with a programmed device such as a safety PLC. For example, a safety valve is not attackable, a microprocessor-based gas detection system is attackable, a mixed human and technical barrier is attackable if it includes a digital device (e.g. a system with an emergency stop button), and a mixed human and technical barrier is not attackable if it is mechanical (valve opening);

7) assessment of severity and likelihood using the method described in section 9.5.1. For a feared event, the maximum value is retained if there are several possible causes;

8) proposal of new safety barriers if needed, either physical or instrumented, and determination of the security level (SL) of the systems which are source of the Cyber-IE and of the systems controlling the instrumented barriers (existing and to be added).

The approach can be seen as a preliminary step for in IT risk analysis (EBIOS or others), as it identifies the security requirements for the industrial control system.

It can also be used as a risk analysis method as part of the approach recommended by IEC 62443 and can be used to determine the target SL of the system. Indeed, a cyber PHA analysis reveals the most important risks, particularly those for which the level of cybersecurity is important, that is, scenarios with a high severity impact and a high probability, not mitigated by other measures.

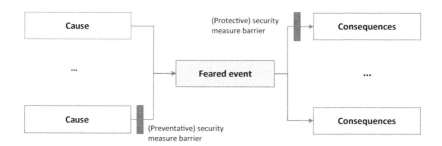

Figure 9.10. *General (appearance) shape of a scenario*

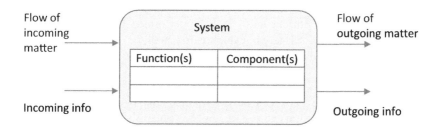

Figure 9.11. *System model*

System	Functions	Components	Input (optional)	Outputs (optional)

Figure 9.12. *Description of the installation*

System	Causes	Derivation	Feared event	Consequences	No barriers		Safety/security barriers	Barriers	
					P|L	S		P|L	S

Figure 9.13. *Cyber PHA analysis table*

For example, in the case of boiler overheating that could lead to an explosion, if temperature control is only performed by a cyber-physical device, the level of SL-target cybersecurity will be high. If there is a mechanical device that stops the heating, the target security level may be lower.

NOTE.– An existing PHA can be transformed into cyber PHA by adding causes of computer origin and reassessing the scenario's likelihood.

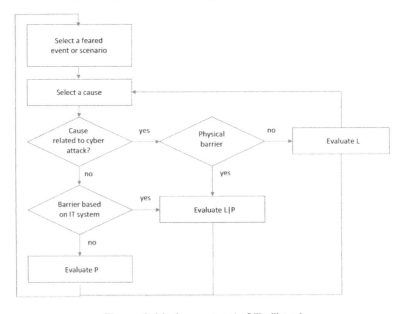

Figure 9.14. *Assessment of likelihood*

9.5.3. Cyber HAZOP

In this section, we present an approach called cyber HAZOP that is an extension of the HAZOP method, which is carried out in a similar way to the extension of the PHA to the cyber PHA method.

This approach should not be confused with the CHAZOP (Computer Hazard and Operability Study) method. CHAZOP is an acronym for control HAZOP/computer HAZOP. The idea of this method is to analyze the effect of deviations of the control system on the evolution of the system. Specific keywords are used and are of the following type: no signal, out of range signal, no power supply, no communication, failure of the I/O board, incorrect or inadequate software programming, cyber-attack, etc.

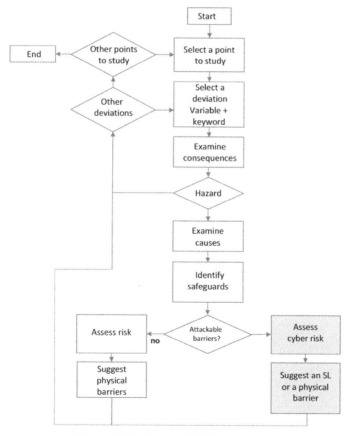

Figure 9.15. *Cyber HAZOP approach*

A cyber HAZOP is a HAZOP analysis (Chapter 8) for which cybersecurity aspects are explicitly addressed. The HAZOP method approach is modified as shown in Figure 9.15. If the effects of the deviation are prevented by at least one non-attackable barrier, such as a safety valve, then the consequences of the deviation cannot be generated by a cyber-attack and is therefore considered non-attackable. In this case, the treatment is classic. Otherwise, if the barriers considered for the deviation are attackable, because they are based on instrumented systems for example. In this case, it is appropriate:

– either to offer a non-attackable barrier with sufficient reliability;

– or to propose a required level of SL based on the organization's risk criteria (Table 9.10).

The choice between the two options will be based on the significance of the consequences and choices for risk management. Typically, for very high consequences, a non-attackable barrier is recommended.

The results of this cyber HAZOP can be used as an input for IT risk analysis. The list of barrier requirements and associated consequences corresponds to security requirements and impacts in the event of loss of control.

Level	Consequences		SL required
1	No discomfort for the stakeholder	No influence on the process	1
2	No discomfort for the stakeholder	Degraded operation	1
3	No discomfort for the stakeholder	Influence on product quality or safety	2
4	Involvement in survival or personal injury	Production stoppage or accident	3
5	Several potential victims	Partial destruction of production equipment	4
6	Many victims off site	Destruction of production site	Physical barrier

Table 9.10. *Example of simplified determination of the SL level based on the consequences*

Process section	Deviation variable + keyword	Causes	Consequences	No barriers		Safety / security barriers	Barriers	
				P|L	S		P|L	S

Figure 9.16. *Cyber HAZOP analysis table*

9.6. Bowtie cyber diagram

Preliminary risk analysis can be supplemented by a detailed analysis: for feared events that may lead to serious consequences, the details of the scenario that generates the feared event are sought. The scenarios obtained by the detailed analysis are represented by a bowtie diagram (Chapter 8).

It is possible to take into account in this detailed analysis the causes of computer attacks. These causes can be due to

– a lack of integrity of the software or control system data;

– a lack of availability, in particular for the SIS.

It is possible to show on the bowtie diagram the causes related to a computer security problem. In the example shown in Figure 9.17, an event that represents corruption of the actuator command (a valve opening that is too small on the diagram) provides the link between the two types of diagrams, and we obtain a cyber-bowtie diagram.

The use of this representation to assess the level of risk must be conducted with caution, as the likelihood of computer attacks and failures have not been of the same order of magnitude. In the first case, it is rather high; in the second case, it represents a probability of failure and is therefore quite low.

The severity level of the damage generated determines the security requirements for the control functions and data involved in the scenario. A dual rating system of failure probability and likelihood of a cyber-attack similar to that of a cyber PHA is recommended (Abdo *et al.* 2018).

In the approach cited, the double rating is composed of a probability, a quantitative variable between 0 and 1 to measure the likelihood of technical and human failures, and a qualitative rating to characterize cyber-attacks, because it is unrealistic and of little use to determine a numerical value.

In a cyber-bowtie diagram, the events linking the two analyses represent a deviation of the input or output process variables. Performing a PHA prior to an EBIOS study is therefore very useful for defining security requirements and prioritizing problems. This will allow for a more detailed analysis of the IT elements that generate the most serious consequences.

9.7. Risk analysis of IIoT systems

Systems using IIoTs have particularities. These systems are designed to add new objects dynamically: the configuration is therefore potentially in a constant state of change. It is therefore difficult to carry out a traditional inventory of the system, and it is necessary to define the perimeter in an abstract way, for example all sensors of a given type that have joined the network and are located in a given perimeter.

The new risks to be taken into account are those related to the addition or removal of the object, and to corruptions already present in the object at the time of its acquisition. The risks associated with the addition are essentially those due to incorrect authentication or incorrect rights being granted at the time of addition. The risks associated with withdrawal are those associated with the disclosure of information leaving the system with the object. Another substantial risk is related to physical manipulation of the object (Chapter 5) to extract information from it.

Finally, another aspect to be taken into account is that IoT objects can be used as an attack platform, especially for objects of the same type; they can be used in parallel using the same process.

Risk Assessment Methods 247

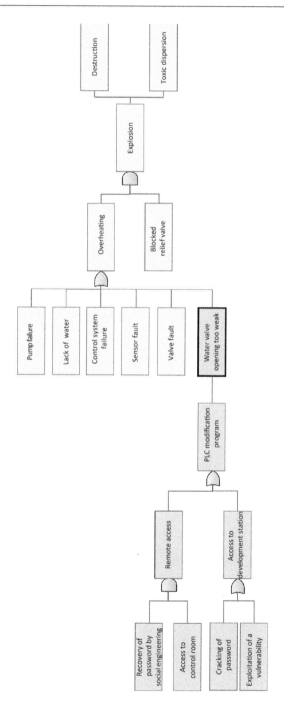

Figure 9.17. *Example of a cyber bowtie diagram*

10

Methods and Tools to Secure ICS

10.1. Identification of assets

A prerequisite for any risk analysis is to carry out an inventory of the installation. This involves identifying the elements of the control system and its interfaces with, on the one hand, the physical world and, on the other hand, the world of data processing. This inventory includes links to the business computing system, Internet access, remote connections by modem or via other types of connections and exchanges via removable memories such as USB sticks.

System components are hardware components, software components and communication equipment. Human elements can be added to it. It is also useful to identify the material and logical links between the machines. We get what is called a mapping of the facility.

If the cybersecurity study takes place as part of the design of a new facility, there are documents describing the functionality and planned implementation. There is also usually a functional safety study on which to base it. In the case of an existing installation, the description of the installation will be based on the actual installation and the various documents available.

First, it is necessary to define the scope of the study: it is chosen to contain all the critical part of the facility or infrastructure (networks, transport, electricity, etc.). If the installation is complex, it may be advisable to divide it into subsystems, if possible by grouping by criticality levels.

The description of the perimeter must be done in terms of functionalities, of geographical location and in terms of time (period of life of the installation in which we are interested). From this perimeter, it must be possible to assess the number of in/outgoing flows from each of a material, informational and human point of view. The next step is the understanding of the general functioning of the installation and the identification of business needs. To do this, we look for the main functions or processes provided by the installation. Then it is necessary to understand the links between the logical (data and software) digital devices and the physical part. The equipment at the interface is mainly programmable logic controllers and IIoT equipment.

Functions or processes	Type	IT or OT components	Physical components or part of the controlled system

Table 10.1. *Description of the system elements*

Functions can be classified according to several types depending on whether they form part of the basic process control system (BPCS), the safety instrumented system (SIS), or whether they are part of the support functions for maintenance, configuration or administration.

Physical components include programmable logic controllers (PLCs), remote I/O, human–machine interface (HMI) units, sensors and actuators, central measuring systems and programming devices. These physical components are supplemented by software entities and data that must also be identified. It is also useful to describe the communications between the various elements and the outside world. Each of the elements must be described in detail, including the following:

– the inventory name;

– the type of equipment: workstation, mobile computer, server, storage unit, switch, gateway, modem, PLC, HMI, etc.;

– the brand and model;

– the version of the operating system;

– the list of applications and their versions;

– the services offered and their versions;

– the version number of the embedded software modules;

– the physical location;

– the list of other devices connected to the different network ports;

– for devices connected to the network, network information (IP address, mask, gateway, MAC address or specific addressing) will also be identified.

This equipment inventory could be supplemented by a certain amount of information on the networks:

– for IP networks: the list of IP address ranges with associated switches and interconnections with other ranges;

– for non-IP networks: the list of MAC addresses of equipment or protocol-specific addresses, associated switches and lists of elements to other networks, in particular PLCs;

– the list of non-Ethernet access points with, for each, the list of access ports, addressing and the list of connected equipment;

– the list of connected servers and workstations;

– the list of programmable logic controllers and equipment connected to the network (remote inputs/outputs, intelligent sensors/actuators, IIoT equipment, etc.).

The communication links between the equipment can be graphically represented (Figure 10.1). It is also possible to represent the network connection diagram, called the "logical view of the installation" (Figure 10.2). This logical view makes it possible to show the different subnetworks, gateways and connections with the outside world.

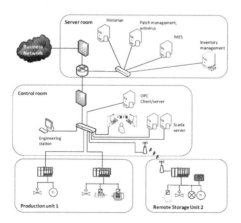

Figure 10.1. *Example of physical mapping. For a color version of this figure, see www.iste.co.uk/flaus/cybersecurity.zip*

252 Cybersecurity of Industrial Systems

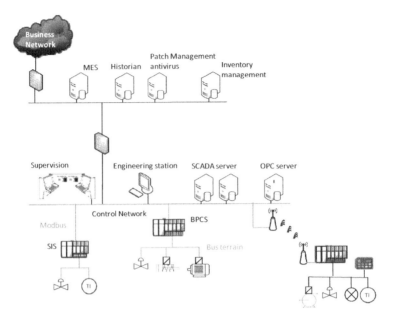

Figure 10.2. *Example of logical mapping. For a color version of this figure, see www.iste.co.uk/flaus/cybersecurity.zip*

After having carried out an inventory of the machines and their connection, it is useful to carry out an inventory of the applications dedicated to ICS and the communication flows between them. These applications include SCADA workstation HMI applications, data historian, PLC programs, etc.

For each application, the following elements can be listed:

– the type of application;

– the owner;

– the number of users;

– the equipment (physical or logical) required for this application to work;

– the services listening on the network and associated ports;

– the data flows between applications;

– the version of the application.

A graphical representation can be made to show the different applications and the flows between them. These data are used to define zones and configure firewalls and other filtering or detection devices presented below.

Inventorying the various workstations, servers and equipment is a task that can be tedious and prone to error. In the IT world, there are tools available that can identify the workstations and running applications. In the OT world, this is not always possible, as equipment operating systems, which may be old or limited in functionality, do not always allow it.

However, a number of network tools can be used, such as arp-scan. In addition, commercial or open source tools with a graphical interface, such as Grassmarlin (Figure 10.3) (NSA 2017), are beginning to emerge.

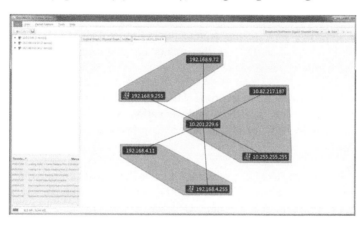

Figure 10.3. *Grassmarlin screenshot. For a color version of this figure, see www.iste.co.uk/flaus/cybersecurity.zip*

10.2. Architecture security

10.2.1. *Presentation*

Most of the networks encountered use the TCP/IP protocol, which also supports industrial protocols such as Modbus, as presented in Chapter 2. Packet routing is based on the Address Resolution Protocol and is relatively easy to bypass. Remember that on a local network, one machine can pretend to be another machine (MitM attack, Chapter 4) and, since the protocols are not secure, it is easy to corrupt the frames exchanged between the equipment. It is therefore very important to isolate as much as possible the industrial control system (ICS) local network from other external networks (IT or Internet in particular).

In the context of an ICS, the design of a secure architecture involves a partioning into zones, with flows between each zone being filtered and monitored. In addition, activity in an area can also be monitored.

10.2.2. *Secure architecture*

Before the 1990s, ICSs were not connected to the entreprise information system network. The risk of computer attacks was controlled through isolation, and physical access or intrusion was required to compromise a system. The convergence of OT and IT has made this approach obsolete.

The objective of a secure architecture is to provide the best possible isolation between IT and OT so that vulnerable protocols and equipment in ICSs are as little exposed as possible.

The basic approach for securing ICS is the defense-in-depth approach. It uses methods such as the implementation of zones and conduits recommended by IEC 62443 to secure communications between trusted spaces.

Purdue's model has been presented in Chapter 2 and serves as the basis for proposing a secure architecture (Figure 10.4). In order to secure exchanges between the IT and OT zones, and to facilitate firewall configuration, a demilitarized zone (DMZ) is often added between levels 3 and 4.

A DMZ is a physical and logical subnetwork that acts as an intermediary between the OT and IT networks. It adds an additional security layer to an organization's local network: IT equipment exchanges with DMZ area equipment, and OT equipment exchanges with DMZ area equipment.

By placing resources accessible from the IT network in a DMZ, no direct communication is required between the OT zone and the IT zone. The firewall can prevent unauthorized packets from the IT network entering the OT network and can also filter traffic from other areas of the network, including the OT network. With a well-designed set of rules, a clear separation between the OT network and other networks with limited or even no traffic between the IT and OT networks can be maintained.

Methods and Tools to Secure ICS 255

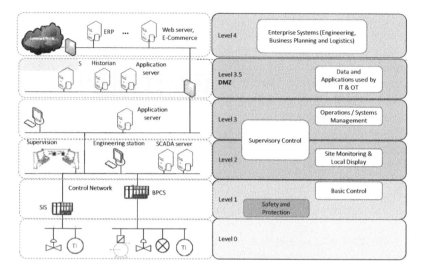

Figure 10.4. *Architecture with a DMZ. For a color version of this figure, see www.iste.co.uk/flaus/cybersecurity.zip*

The main security risk in this type of architecture is the compromise of a computer in the DMZ that is used to launch an attack against the OT network via authorized application traffic from the DMZ to the OT network. This risk can be reduced by regularly hardening and updating servers in the DMZ, and by configuring the firewall with rules that only allow connections initiated by OT network devices to the DMZ.

10.2.3. *Partitioning into zones*

A security zone is a physical and/or logical grouping of resources or grouping of resources with similar security requirements. This concept is defined in particular in IEC 62443 (Chapter 7). Physical zones are defined by grouping resources according to their physical location. Logical or virtual zones are defined by grouping resources according to functional criteria or by analyzing communications with a given protocol. To define the zones, the physical and functional model of the control architecture (mapping) developed during the asset identification phase is used.

A security policy for each area must be defined. It is possible to define areas within areas, which makes it possible to set up a defense in depth.

Information must be able to flow within a security zone and from a security zone to the outside and, conversely, from the outside to the inside. Even in a non-networked system, some communication exists (for example, intermittent connection of programming devices to create and maintain systems or transport of information via USB sticks). These communications to and from the outside define the entry and exit points of a zone. The resources used for communication can be grouped into a specific zone, called a "conduit".

A conduit can be a single service (i.e. a single Ethernet network) or can consist of multiple data carriers (multiple network cables and direct physical access). As with zones, they can be physical or logical constructions. Conduits can connect entities in a zone or connect different zones.

The structuring of a system into zones and conduits makes it possible to define a level of security suited to each zone and to set up devices at the level of the conduits to limit propagation of the effects of a feared event in the same way as the watertight compartments of a ship, and then to set up a defense in depth by interlocking the zones and the conduits.

The zone concept goes beyond Purdue's reference model. This model has been proposed to organize the integration of IT and OT. The division into zones and conduits has a different objective of structuring the system in relation to security requirements.

To define the zones, we start from the physical diagram of the architecture, organized if possible according to the Purdue model. A first level of partitioning is the breakdown according to levels.

The rest of the breakdown is carried out by analyzing the functionalities of the different elements and the protocols used. The different phases from operation to maintenance and configuration are reviewed. Subfields can be created if necessary. In addition, some resources are grouped into specific zones: safety instrumented systems (SIS), wireless equipment, temporarily connected equipment (programming console) and equipment connected via unsecured networks such as remote equipment.

When a resource supports several functions, it will be assigned to an area corresponding to the most restrictive function requirement, or a separate area will be created with a specific security policy. If, for example, several applications involving different activity levels are running on a single physical server, a logical zone can be created. In this case, access to a particular application is limited to users with privileges for that application level. An example is a single machine running an OPC server and analysis tools based on the OPC client. Access to the OPC server is limited to users with higher level privileges, while access to spreadsheets using the OPC client plug-in is available to all users.

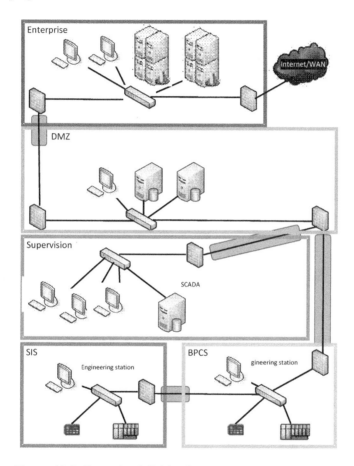

Figure 10.5. *Example of division into zones and conduits. For a color version of this figure, see www.iste.co.uk/flaus/cybersecurity.zip*

10.3. Firewall

A firewall is a device that allows the partitioning policy to be applied between two zones. It makes it possible to filter the data flows, to monitor them and to log them. There are two types of firewalls: network firewalls and firewalls installed on hosts, for example, related to the operating system. The operating principles are identical, but securing the architecture is mainly based on network firewalls.

The fundamental technical function of any network firewall is to filter Transmission Control Protocol (TCP)/Internet Protocol (IP) communication packets. It is configured by defining rules, which can use source or destination addresses or ports. The firewall inspects each packet it receives to determine if it meets the rules. For example, traffic that does not come from a defined source can be blocked. A firewall is used at the border of an area. Rules can, for example, limit traffic to a single server, or allow it only with certain types of workstations, such as maintenance workstations.

The security provided by a firewall is closely linked to the accuracy of configuration rules. In general, for IP flows, identified by source IP address, destination IP address, protocol (e.g. UDP or TCP) and, if applicable, source and destination port numbers that are related to the type of application, the principles are as follows:

– anything that is not explicitly allowed is prohibited or, in other words, flows are refused by default;

– only the flows necessary for the functioning of the industrial system are allowed;

– rejected flows must be logged and analyzed;

– all incoming and outgoing flows in the industrial system must be logged.

In the case of an industrial system, exchanges between IT and OT parts must be kept to a minimum.

When a DMZ is used, it is possible to configure the system so that no traffic is allowed between the two zones: it ends up on the servers of the DMZ zone. For example, IT network workstations communicate with the DMZ historian server using an http protocol, whereas OT network equipment communicates with this server using Modbus.

This approach allows a protocol break and significantly increases security. Table 10.2 summarizes the configuration rules. It should be noted that rules 9–11 concern the security of the firewall itself, which should not be neglected.

In the case of industrial systems, there are firewalls that inspect the contents of the packet and are able to recognize industrial protocols. In addition to analyzing source and destination addresses and port numbers, the firewall analyzes the content of the frame. It is, for example, able to decode the Modbus protocol and recognize the type of function, the address written or read and the data sent. This is called Deep Protocol Inspection (DPI). From the configuration of the PLCs and the SCADA, it is possible to define rules on these elements. For example, rules can block writing to a memory area, or limit the values of a command to a maximum threshold. A number of commercial products implement these features, as well as open source software such as Snort, Suricata or Bro presented below.

No.	Rules and regulations
1	All the basic rules must be "block everything".
2	All authorized flows must be specific to the source and destination address, as well as port numbers, and stateful if necessary.
3	Any direct traffic between the control network and the corporate network should preferably be blocked. All traffic should end in the demilitarized zone.
4	Flows between the control network and the corporate network can be authorized on a case-by-case basis, after risk analysis.
5	Any protocol allowed between the control network and the demilitarized zone should not be explicitly allowed between the demilitarized zone and the corporate networks (and vice versa) in order to ensure a protocol break.
6	All outgoing traffic from the corporate network to the control network should be restricted by source, destination and port filtering.
7	Outgoing packets from the control network or demilitarized zone should only be allowed if they have a correct source IP address that is assigned to the control network or devices in the demilitarized zone.
8	The devices of the control network must not be directly connected to the Internet, even if the area is protected by a firewall.
9	All firewall management traffic must be carried out either on a separate secure management network or on an encrypted network with multifactor authentication. Traffic should also be limited by IP address to specific management stations.
10	All firewall policies should be tested periodically.
11	All firewall configurations must be backed up immediately before commissioning.

Table 10.2. *Configuration rules*

Remote access is an important issue for firewall rule management. It is very common for remote access to be required for maintenance by suppliers or integrators, or even for remote operation. Access must be managed with an appropriate authentication system, usually multifactor, and a secure network, such as a virtual private network (VPN). The access management server must be placed in the DMZ or be part of the IT infrastructure. It may be useful to set up a second level of authentication to access the OT network.

10.4. Data diode

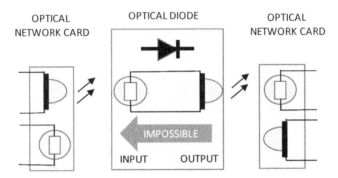

Figure 10.6. *Principle of a data diode. For a color version of this figure, see www.iste.co.uk/flaus/cybersecurity.zip*

A data diode, also called a "unidirectional gateway", is a device that allows data to move only in one direction. The system can be software or hardware, for example based on an optical system, which guarantees the blocking in the opposite direction.

A data diode is required for Class 3 facilities in the ANSSI classification. A data diode can be coupled to a software device to replicate databases and emulate a server.

A data diode can be used to let the data pass only in the OT to IT direction to upload production data. Bidirectional communication can be achieved using multiple data diodes and a DMZ with protocol break, which is well suited for securing IT/OT connections.

Methods and Tools to Secure ICS   261

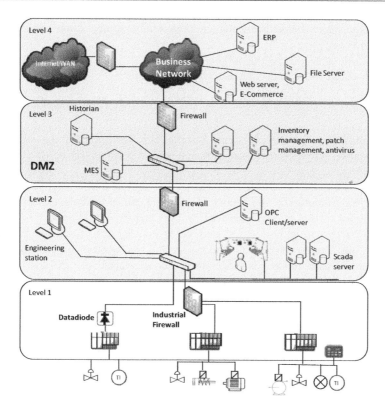

Figure 10.7. *Example of an architecture with data diode, firewall and industrial firewall. For a color version of this figure, see www.iste.co.uk/flaus/cybersecurity.zip*

10.5. Intrusion detection system

10.5.1. *Principle of operation*

Intrusion detection consists of monitoring events occurring in a computer system or network, and analyzing them in order to detect possible incidents that constitute violations of security rules or policies, or imminent threats to the information system or installation.

Incidents have many causes, such as malware (e.g. viruses or spyware), malicious users accessing systems from the Internet and authorized users of systems who abuse their privileges or attempt to obtain additional privileges.

Although many incidents are malicious in nature, some are not and may be human errors.

There are two types of intrusion detection systems:

– Network intrusion detection systems (NIDS) that monitor network flows;

– intrusion detection systems installed on workstations that monitor workstation activity (host-based intrusion detection system [HIDS]).

A NIDS collects network traffic using one or more probes, and analyzes it according to protocol, type, source, destination, etc. It is deployed to be able to monitor traffic entering and leaving the network area (Figure 10.8), and looks for certain activities that may be hostile actions or misuse, such as denial of service attacks, port scans, malicious content in data packets (Trojan, viruses or worms, etc.), brute force attacks, etc. An NIDS can eventually block traffic: this is called an intrusion protection system (IPS).

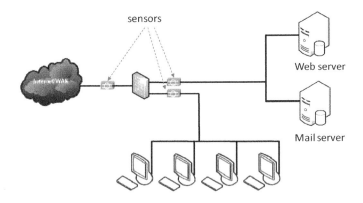

Figure 10.8. *NIDS. For a color version of this figure, see www.iste.co.uk/flaus/cybersecurity.zip*

The advantage of a NIDS is to provide intrusion detection coverage with fewer system resources and with generally lower deployment, maintenance and update costs than a HIDS. In addition, it has visibility over all network traffic, and it can correlate attacks between several systems. The disadvantages are related to the need to set up a large number of probes, a secure network to collect information and a relatively high processing capacity to avoid slowing down operation.

Another limitation to be aware of is that a NIDS is ineffective when traffic is encrypted.

Intrusion detection systems can also be deployed on hosts, workstations or servers (Figure 10.9). This implementation is called HIDS. A HIDS examines log files and looks for events such as connections at strange times, connection authentication failures, addition of new user accounts, changes or access to critical system files, changes or deletions of binary (executable) files, starts or stops of particular processes, attempts to increase privileges or use certain programs. It also examines network traffic entering or leaving a specific host, with some also supporting specific applications (FTP or web services).

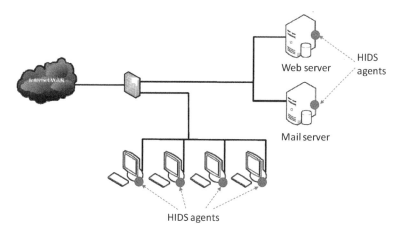

Figure 10.9. *HIDS. For a color version of this figure, see http://www.iste.co.uk/flaus/cybersecurity.zip*

These systems are beneficial but heavy to deploy and less well suited to ICS, where the main focus is on monitoring network traffic, and where there are many devices on which to deploy HIDS is not possible. In many cases, a combination of both approaches is used.

The general structure of the processing modules of an intrusion detection system (IDS) is shown in Figure 10.10. The data come from network traffic or host activity. The analysis engine uses these data and knowledge about the attack profile or normal system behavior to detect an intrusion. The alerts generated can be transmitted to a Security Information and Event Management (SIEM) that provides real-time analysis of security alerts generated by applications and network equipment. This system is described in the following section.

IDSs are systems used for monitoring, which is the third function of the NIST cybersecurity framework presented in Chapter 11.

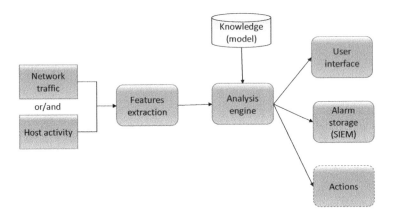

Figure 10.10. *Structure of an IDS. For a color version of this figure, see http://www.iste.co.uk/flaus/cybersecurity.zip*

10.5.2. *Detection methods*

To detect incidents, an IDS uses knowledge or a model, which can represent normal or abnormal behavior:

– in the first case, the model describes the normal behavior and detection is based on the search for anomalies in relation to this behavior; we speak of an anomaly-based system;

– in the second case, the model describes behaviors characteristic of an intrusion and is referred to as a signature-based method.

Models are built by experts or by automatic machine learning methods.

If the model is inadequate, for example if the signature is missing or inappropriate, or if normal behavior is not captured correctly, then the IDS can generate an alarm for benign traffic that is not hostile, this is called a false positive. Similarly, hostile activity, which does not match an IDS signature or resembles normal activity, may be undetected, this is called a false negative (false sense of security).

This imperfect functioning is an important limitation to the use of IDS because it can be perceived as a system generating disturbances due to false positives. This is also the reason why IPS mode operation is almost never used, as the risk of interrupt normal operation is too great. This is especially true for industrial systems.

Detection approaches are presented in more detail below. Many IDSs use both approaches, either separately or in an integrated way, to improve performance.

10.5.2.1. *Signature-based approach*

The idea is to describe what is known to be representative of an attack by a model called a signature, and to monitor network traffic to find matches to these signatures. There are signatures:

– based on content such as, for example, character sequences such as/etc./passwd in a telnet session;

– context-based, such as a signature that corresponds to a potential intruder performing a scan to find open web servers, or a signature that identifies a Nessus scan, or a ping flood attack.

Signature-based detection is very effective in detecting known threats, but it is ineffective in detecting unknown threats or many known threat variants. Signature-based detection technologies have a poor understanding of many network or application protocols, and cannot track and understand the status of complex communications. For example, signature-based IDSs cannot associate a request with the corresponding response, such as knowing that a request to a web server for a particular page has generated a response status code of 403, which means that the server has refused to respond to the request. Nor do they have the ability to remember previous requests when processing the current request. This limitation prevents signature-based detection methods from detecting attacks that include multiple events, if none of the events contains a clear indication of an attack.

In addition, to apply this approach in the case of an ICS, the attack signature database must have different characteristics from those of signatures in an IT-oriented database. In particular, signatures must take into account industrial protocols.

IDS and IPS providers develop and incorporate attack signatures for various ICS protocols such as Modbus, DNP3 and IEC 60870.

10.5.2.2. Anomaly-based approach

Anomaly-based detection consists of comparing the observed activity with a model of the activity considered normal. An IDS using anomaly-based detection has profiles or models that can represent the normal behavior of users, hosts, network connections or applications.

The first question to ask in order to implement such an approach concerns selection of the characteristic elements of the activity. These may include raw network data, protocol behavior (state), substation activity indicators, or even cyber-physical system behavior characteristics. The latter possibility, detection of anomalies with respect to the cyber-physical system model, is discussed in the following section.

The second question concerns the methodology for constructing the model: specification by the analyst or automatic learning and, in the latter case, choosing the learning method (Flaus and Georgakis 2018b), which can be a statistical or deep learning approach for example, and then defining the periods during which the activity is considered normal and representative.

Among the advantages of anomaly-based detection approaches are, first of all, their adaptability to both unknown and internal attacks. For example, it is possible to quickly detect an internal attack (using a compromised user account, for example), because the behavior will be unusual. In addition, an attacker never knows what may or may not generate an alarm, making detection more difficult to circumvent.

Disadvantages include the need for a high initial preparation time, the fact that there is no protection during initial learning, the difficulty of characterizing normal behavior and the interpretation of false positives or false negatives, which can be difficult. This approach is still not widely used in commercial products.

10.5.2.3. Practical tools

There are IDS software programs available that can analyze the traffic of industrial systems such as Snort, Suricata or Bro. Snort rules have been developed for Modbus TCP, DNP3 and IEC 60870. Snort is an open-source intrusion detection and prevention system that uses rule-based language to perform inspections based on signatures, protocols and anomalies. Rules for DNP3 and Modbus protocols have also been added to the Bro IDS platform.

Ossec is an open-source HIDS system that performs log analysis, file integrity analysis and many other parameters.

10.5.3. *Intrusion detection based on a process model*

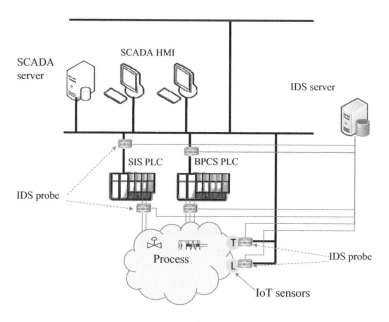

Figure 10.11. *Structure of an industrial IDS. For a color version of this figure, see http://www.iste.co.uk/flaus/cybersecurity.zip*

For cyber-physical systems, in particular ICS, there is a specific detection approach based on the detection of anomalies in relation to the behavior of the physical system model or the control system.

This method is based on the assumption that an attacker's objective is to put the target system in a critical state, so by monitoring the evolution of the cyber-physical system and detecting the occurrence of critical or precritical states, it may be possible to detect attacks. These techniques are based on a model of the physical system and the control system. They are based on ideas similar to those developed for monitoring and diagnosing physical systems.

This approach has developed in recent years, and applications have been proposed in different fields such as chemistry (Cárdenas *et al.* 2011; Fovino *et al.* 2012), water treatment (Hadziosmanovic *et al.* 2014), electricity distribution (Liu *et al.* 2011) and even medical systems (Hei *et al.* 2013).

From an operational point of view, the following elements are necessary to monitor and analyze the evolution of a system:

– a representation language to formally describe the physical system and critical states;

– a system for monitoring or reconstructing the current status of the monitored system;

– a critical state distance metric to calculate the proximity of a state to critical states;

– a detection and possibly prognostic algorithm to decide whether or not the system is evolving toward a critical state.

Different possibilities exist to implement these features. The difficulties encountered with this type of approach are the need to develop a model of the system, which must be robust, and the management of uncertainty regarding the gap between the modeled behavior and the actual behavior. The first point is a limitation that is all the more important as systems often evolve and the model must therefore be readjusted. Dedicated automatic learning can be a solution (Patents, Université Grenoble-Alpes 2018).

10.6. Security incident and event monitoring

The various devices presented in the previous sections (firewall, IDS, data diode), workstations, servers, network equipment, ICS equipment, authentication systems, generate events that characterize the system status and incidents that occur. This information includes:

– information on blocked traffic provided by firewalls;

– intrusion detection alerts provided by IDSs;

– unsuccessful login alerts from authentication systems;

– information about identified malware provided by antivirus software.

The centralization and exploitation of this information is useful for monitoring the security of an ICS. This is the role of SIEM.

The functionalities of these systems are as follows:

– generation of centralized views, which are generally presented in the form of dashboards;

– normalization of received events, on the one hand, by translating them into a standard and readable format, and, on the other hand, by grouping them according to the categories defined by the configuration;

– correlation of events: from predefined rules, it is possible to generate alerts. This can be done on historical data or in real time. An example of rules can be "if more than five unsuccessful logins in 5 min then generate an alert";

– reporting and alert: different ratios can be calculated, evolution curves can be plotted and, from the sequence of pasted events and depending on the ratios, alerts can be generated;

– log management: this allows events and logs to be stored in a central database, while ensuring the quality of storage and the compliance of storage conditions to requirements.

In addition, it may be useful to compare the observed situations to during a situation defined as being "normal". To do this, a SIEM can be used to define a baseline behavior.

There are a number of open-source tools available to implement a SIEM. Examples include the ELK stack or the OSSIM software.

Figure 10.12. *SIEM screen. For a color version of this figure, see http://www.iste.co.uk/flaus/cybersecurity.zip*

10.7. Secure element

IIoT equipment has particular vulnerabilities compared to conventional equipment such as PLCs:

– from a material point of view, they are more easily accessible and can be the subject of physical attacks such as auxiliary channel attacks based on power consumption analysis (Chapter 4);

– from a software point of view, they are programmed using traditional computer languages, less constrained than the languages used on PLCs, which can lead to the creation of software with significant vulnerabilities;

– they are based on generic elements (COTS), which can potentially be corrupted in the supply chain and, for example, contain backdoors.

Systems are not frozen, which poses the problem of adding the device to the network and removing them.

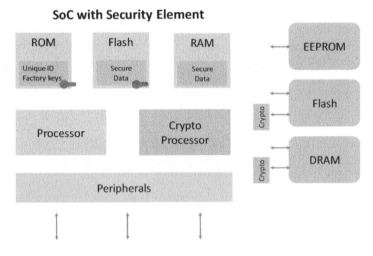

Figure 10.13. *SoC with security element. For a color version of this figure, see http://www.iste.co.uk/flaus/cybersecurity.zip*

In order to secure a system with an IIoT device, it is necessary to carry out the following:

– the authentication of the device;

– the security of data transfers;

– securing the storage of sensitive data on the device;

– provide a secure code execution environment.

The solution is to use a secure element (SE), defined as a tamper-proof platform (usually a secure microcontroller on a chip), capable of securely hosting applications, and storing their confidential and cryptographic data in accordance with the security rules and requirements defined by a set of clearly identified trusted authorities.

A SE is used, for example, to store a device's private key as well as security certificates. To be as safe as possible, it is preferable that it be integrated into the System-on-a-Chip (SoC) itself (Figure 10.13). Indeed, the use of an external circuit is vulnerable to auxiliary channel attacks by observing the transit between the SE and the main circuit.

A secure SoC provides some or all of the following features:

– secure key storage: the keys can be stored in unmodifiable read only memory (ROM) memory at the time of manufacture or in flash memory during configuration. The storage must resist physical attacks (side channel attacks, for example);

– secure boot: the boot code must also be securely stored in a non-editable memory (ROM). This code is considered secure (root of trust). On power-up, the processor starts executing the boot code. Its main task is to start the application code after verifying its signature (Appendix 1). This allows a secure boot;

– secure firmware/software update: the update of the device is carried out in a secure way using signature and, possibly, encryption mechanisms for confidential data;

– Data at Rest Protection (DARP): in the event of physical access to the device, the data must not be readable. The device must therefore be able to encrypt the data stored in flash memory or RAM;

– crypto-processor: in order to be able to easily implement data encryption, it is desirable that it be implemented by hardware functions;

– Trusted Execution Environment (TEE): a TEE is a set of mechanisms at the main processor level that provide secure isolation of libraries and sensitive data. This secure environment ensures that data are stored, processed and executed as planned (Bar-El 2010). The purpose is to avoid any attack on this code, or that sensitive keys or secret data are not read;

– true random generation number: an electronic circuit is used to produce encryption keys that are impossible to predict. The generation of random numbers is useful for cryptographic algorithms;

– unique identifier: it is essential to be able to identify the device in a secure and unique way. Storing an identifier designed to be unique and secure in the SoC is an ideal solution.

In addition, it is desirable that the system should have functionalities to facilitate safety management. Indeed, when a large number of devices are deployed, a centralized safety management system is required. The latter interrogates each device, which must be able to provide adequate information on its characteristics, its condition and any security events.

Depending on the functionalities implemented, the SoC can be used to meet the basic, enhanced or critical levels defined by the Industrial Internet Consortium's good practices (Chapter 6, section 6.9).

11

Implementation of the ICS Cybersecurity Management Approach

11.1. Introduction

11.1.1. *Organization of the process*

Securing a control command system has many facets. Several possibilities exist to organize the process. A fairly natural approach is to consider the different phases of the risk evolution, or more precisely the course of events leading to damage.

First of all, it is necessary to put in place controls to limit the possibility of occurrence of unwanted events. In the case that, despite these controls, attacks occur, it is important to be able to detect them and provide appropriate mechanisms. If an attack is detected, to limit damage, an appropriate response must be planned and implemented. Finally, after this crisis phase, if damage has occurred, we must be able to restore the system and learn lessons to improve system security. This approach is the one proposed in the well-known NIST Cyber Security Framework (Figure 11.1), which focuses on the different phases of risk management:

– identification;

– protection;

– detection, during operation;

– the response to attack;

– recovery.

For each phase, a number of areas have been identified. They include a list of security measures to be implemented.

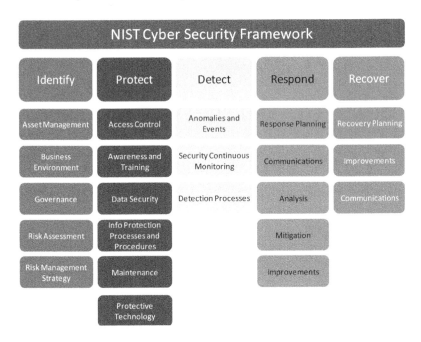

Figure 11.1. *Structure of the NIST framework. For a color version of this figure, see www.iste.co.uk/flaus/cybersecurity.zip*

For the protection phase, an approach based on the defense-in-depth approach makes it possible to properly structure actions in the case of industrial systems. It is very similar to the LOPA method, used for the operational safety of industrial systems, or the defense-in-depth approach used in nuclear power.

A partitioning into zones and conduits as proposed by IEC 62443 (Chapter 7) makes it possible to set up a defense in depth in an appropriate way (Figure 11.2).

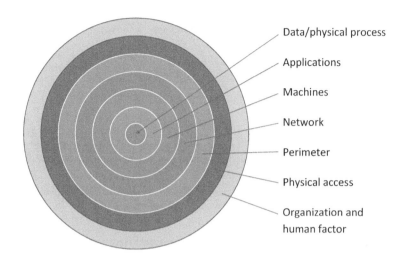

Figure 11.2. *Defense in depth. For a color version of this figure, see www.iste.co.uk/flaus/cybersecurity.zip*

11.1.2. *Technical, human and organizational aspects*

Risk analysis approaches make it possible, either by using a checklist or by using a detailed approach, to identify the vulnerabilities of a system and the security measures to implement in order to achieve a given level of control. It is important to remember that a significant number of these measures are not technical and concern the organization and the human factor. Many of them are of course not specific to industrial control system (ICS) and concern the information system as a whole.

11.1.3. *Different levels of implementation and maturity*

It is not always possible to implement all measures at once. Most often, the measurements are organized according to the diagram in Figure 11.3. The protection measures are the first to be implemented, then the monitoring systems are deployed in the second phase.

In addition, the measures put in place can be more or less well controlled. Quality of implementation is assessed using the level of maturity, such as that proposed by the IEC 62443 approach (Chapter 7), inspired by the CMMI scale.

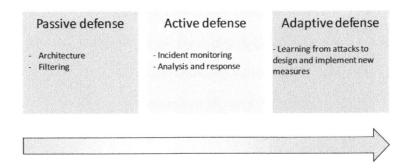

Figure 11.3. *Different levels of countermeasures. For a color version of this figure, see www.iste.co.uk/flaus/cybersecurity.zip*

The rest of this chapter first proposes a simplified approach to securing ICS, and then presents all the steps of a more detailed approach. The choice between the two types of approaches depends on the issues and the level of risk.

11.2. Simplified process

If the stakes are low, a simplified approach to risk control may be sufficient. It is based on the 13 good practices proposed by ANSSI (ANSSI 2012a), which cover all important aspects.

It is useful to have carried out a mapping of the equipment (section 11.4.1), even if it is simplified, to determine:

– the workstations and servers in the perimeter;

– network equipment;

– PLCs and other low level control-command devices.

Based on the description of the installation and knowledge of its scope, the approach, which is part of a defense-in-depth approach, consists of implementing 13 good practices, described in more detail later in this chapter:

– BP01: physical access control to equipment and fieldbuses (Chapter 1, section 1.1);

– BP02: partitioning networks (section 11.10);

– BP03: ensure that off-network exchanges via removable media are also partitioned (section 11.11);

– BP04: user access control and user authorization management (section 11.14);

– BP05: harden the configurations (section 11.12);

– BP06: set up a logging of abnormal events and generate alarms (section 11.16.1);

– BP07: manage PLC configurations and programs to avoid unwanted changes (section 11.13);

– BP08: make validated backups (section 11.19.1);

– BP09: have up-to-date documentation on the installation, reserved for the authorized users (section 11.4.2);

– BP10: implement antivirus protection with regular updates (section 11.12.6);

– BP11: apply security patches regularly (section 11.17.2);

– BP12: protect the PLCs as well as possible given the technical possibilities (section 11.12.3);

– BP13: pay particular attention to engineering stations, ensuring that the good practices mentioned above are properly applied, especially for mobile consoles (section 11.12.2).

These measures are all detailed below. They provide a satisfactory level of security for a non-critical installation.

11.3. Detailed approach

The above pragmatic approach consists of applying a number of relatively generic measures, which make it possible to obtain a "good" level of security, and to maintain it at a minimum by setting up regular updates and monitoring.

If we wish to implement an approach that is more adapted to the installation, and with a more detailed understanding of the measures to be implemented, the approach begins with a risk assessment, and includes a number of steps presented in Figure 11.4. They can be classified into five categories.

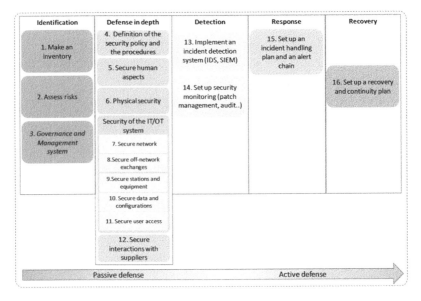

Figure 11.4. *Steps in the ICS security process. For a color version of this figure, see www.iste.co.uk/flaus/cybersecurity.zip*

– *Identification*:

1) inventory the facility (section 11.4);

2) assess the risk (section 11.5);

3) set up a governance system and, optionally at first, a risk management system (section 11.6).

– *Protection (with defense in depth)*:

4) secure organizational aspects with the definition of policy and procedures (section 11.7);

5) secure human aspects: awareness, training, training (section 11.8);

6) secure from a physical point of view (section I.1);

7) secure the network (section 11.10);

8) secure off-network exchanges (section 11.11);

9) secure stations and equipment (section 11.12);

10) secure data and configurations (section 11.13);

11) secure user access (section 11.14);

12) secure interactions with suppliers (section 11.15).

– *Detection*:

13) implement an incident detection system (section 11.16);

14) set up security monitoring (section 11.17).

– *Response:*

15) *establish* an incident handling plan and an alert chain (section 11.18).

– *Recovery:*

16) *implement* a recovery and activity continuity plan (section 11.19).

These different steps are detailed in the following.

11.4. Inventory of assets

11.4.1. *Mapping*

In order to be able to assess the risks, or even simply to systematically implement good practices, it is necessary to first make an inventory of the various elements of the installation.

This approach was described in Chapter 10 (section 10.1).

It makes it possible to obtain a physical mapping and a logical data flow mapping.

11.4.2. *Documentation management*

The description of the mapping, the technical documentation of the installations, the architecture and geographical location diagrams and the addressing plan provide an image of the installation. These documents can be supplemented by user manuals, maintenance plans and operating documents. Some may contain passwords (such as some on-call documents).

All these documents must be managed to be as up-to-date as possible and also secured, as they provide sensitive information to potential attackers.

A documentation management policy (update process, retention period, mailing list, storage, etc.) must be defined. Documentation relating to an information system should not be kept on the system itself.

11.5. Risk assessment

The objective of this step is to assess the risk level of the installation. Risk assessment methods are presented in Chapter 9. The question is:

– on the one hand, what the impacts of a cyber-attack may be, particularly in terms of material damage to people or property, or on production capacities;

– on the other hand, what is the likelihood of occurrence of the undesirable events that caused these impacts.

This likelihood may depend on the vulnerability of the ICS, but also on the operating security measures to protect against poor control system behaviour.

The objective of this analysis is to identify the most critical parts of the installation and, depending on this criticality, to decide on the different security measures to be implemented.

A question for risk analysis is whether the installation can be broken down into different zones *a priori*, which is not necessarily easy. The IEC 62443 standard proposes carrying out a preliminary global analysis to facilitate this breakdown into zones, then to proceed to a detailed analysis of each zone.

To perform this risk analysis, either the risk analysis methods presented in Chapter 9 or the ANSSI class approach (Chapter 6), which allows the criticality level to be determined in a simplified way, may be used.

11.6. Governance and ISMS

11.6.1. *Governance of the ICS and its enviroment*

One of the challenges of ICS security management is that Operation Technology (OT) security governance and Information Technology (IT) security governance are generally managed by different managers, with different cultures and objectives.

OT equipment is managed by the operating manager or automation department, while IT equipment of the ICS is managed by the IT department and, in some cases, by both parties (Figure 11.5). Often, there is doubt about who has responsibility for the security of the ICS infrastructure, which can lead to serious gaps in security management.

The various responsibilities and missions must therefore be analyzed and documented. This information will be used to define roles and responsibilities in the security policy.

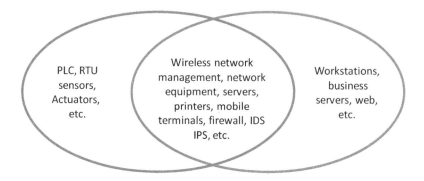

Figure 11.5. *OT/IT responsibilities*

11.6.2. *ISMS for ICS*

The approach presented in section 11.3 describes the steps to implement an ICS security approach. Ultimately, this approach must be part of a PDCA-type security management system (Chapter 3). Steps 1–13, "Identification", "Protection" and "Incident detection" correspond to the *plan* and *do* phases. The *check* phase, which consists of monitoring security, is partially addressed in step

14 (audit). In this phase, two types of indicators are being monitored to determine:

– if the security measure is (correctly) implemented by the organization, such as the deployment rate of security patches;

– if the security measure effectively reduces risk, for example, by measuring the number of contaminated workstations or the number of users clicking on a hacked email (preferably sent by the IT department for testing purposes).

This information makes it possible to correct the security measures implemented at each of the steps proposed in Figure 11.4 in the *act* phase.

11.7. Definition of the security policy and procedures

From an organizational point of view, the first step is to define security policies and procedures for OT-related aspects in order to complement the company's ISSP (Information Systems Security Policy). It must be written in, taking into account the compliance with laws and regulations.

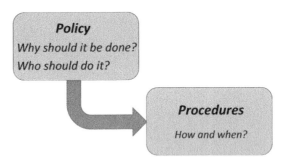

Figure 11.6. *Security policy and procedures*

The ICS security policy can begin by describing the context and scope of the ICS and then recall the challenges of industrial security. It then defines the ICS security organization with the definition of roles and responsibilities, in particular between the IT and OT world, as well as relations with external service providers. In the case of an operator of essential service (OES), the aspects concerning relations with the authorities must be defined.

The security policy must also address issues such as user account management, use of removable media, remote access, mobile devices, use of personal terminals, antivirus management, security update management, change management, backup and recovery procedures and incident response. It can also deal with documentation management (update process, storage, duration), outsourcing management and clauses to be included in contracts.

It is recommended to produce documents independent of the ISSP of IT systems and to refer to them if necessary.

Based on the security policy, the procedures and guidelines implementing the policy are developed. It is of course possible to rely on the recommendations of the various guides and standards (Chapter 6). The guidelines define the rules of use, good practices, etc. They may concern, for example, the operating operator, the network administrator or external service providers. Procedures describe how to perform the different tasks, for example, how to perform the vulnerability analysis in a concrete way.

11.8. Securing human aspects

The human factor is important in risk management. Training or awareness-raising, possibly supplemented by situational exercises or tests, must be carried out. This training, which is still often neglected in the OT world, must be adapted to the types of tasks and responsibilities of each individual. We will distinguish for example:

– operating personnel;

– managers;

– personnel performing maintenance or development;

– new hires;

– external service providers;

– the visitors.

Training should ensure that staff are aware of the security policy and procedures and implement them. The first step is to develop an awareness program for all employees involved. This awareness must be supported by management. It must then be followed by regular communications to remind us of the risks and important points.

In a second step, a more targeted training program should be conducted. It is recommended to base it on the roles for the security of ICS. Designing a role-based training program begins by identifying the main roles; this identification can be based on the list presented above (operating staff, managers, etc.). Then, training needs are identified for each role. For example, visitor training can focus on defining authorized and prohibited activities on site, while development staff training can focus on configuration management issues and the secure use of network resources. Management training may focus on how to react when an employee reports a possible security violation.

11.9. Physical security

Physical access to equipment is one of the potential vectors of attack. It is possible to insert a USB key, set up a keylogger, connect directly to the network, steal data or even equipment for reverse engineering purposes, or simply destroy machines.

Physical security is therefore very important, as is logical access control.

The approach to be implemented, similar to the security of logical access, consists of defining the perimeter to be protected, identifying the roles of the various stakeholders, dividing the installation into zones and defining accesses to the zones depending on the roles. Connection cables must also be protected, as well as remote equipment.

Several layers of protection can be implemented, such as a fence around the facility, and then locked doors on the ICS building, as well as additional locked doors for the control room and equipment rooms.

Access can be authorized using different types of techniques (badge, keys, etc.), monitored with cameras and connected to a logging system connected to the SCADA supervision.

Access management based on roles and staff taking into account new employees and employment termination must be provided.

The access control system must be compatible with management of emergencies related to physical risks (e.g. fire, toxic leak). A glass breakage type system can be used.

11.10. Network security

Network security consists of setting up a secure architecture and then performing perimeter monitoring (firewall or data diode). The methodology and systems for filtering data flows are described in Chapter 10.

Access to the network is authorized according to roles and users. This requires an access control system as described in section 11.14.

Remember that a virtual network is not a risk-free solution: the source and destination are vulnerable; if malware is installed on the source computer, it can infect the destination computer without even being detected, since the flows are encrypted. This issue is addressed in Chapter 10.

11.11. Securing exchanges by removable media

A policy for the use of this type of media must be defined. At a minimum, the software restrictions (no autostart from removable media) must be activated.

For greater security, these media should not be used, and transfers should be made through a dedicated and secure workstation with antivirus software.

11.12. Securing machines

11.12.1. *Securing workstations and servers*

The various equipment, computer workstations, network equipment, servers, PLCs and specific equipment must be secured with measures that reduce vulnerability such as:

– the deactivation of default accounts;

– closing open ports and using non-standard ports (for example, for SSH, do not use port 22);

– uninstalling unused applications and services such as test tools, debugging tools or network functions;

– the periodic installation of security patches (section 11.17.2);

– the implementation of a "whitelisting" system for applications (authorized list).

A host-based intrusion detection system (HIDS, Chapter 10) can be set up and connected to the Security Information and Event Management (SIEM).

11.12.2. Securing engineering stations

Engineering stations are special workstations. They are used for development and to send the code to the PLCs. They therefore contain critical information and can, because of the development environments that are present, offer more vulnerabilities. Analysis of attacks carried out shows that they are often a vector of attack, as this was the case for Stuxnet.

In addition to the general recommendations, it is necessary for these workstations:

– systematically to identify and authenticate users;

– to check that the antivirus update is functional;

– to check that the OS and software are updated;

– not to connect mobile engineering workstations to networks other than the OT network;

– to shut down the workstations when they are not in use.

Firewalls must be configured so that these workstations can be updated.

11.12.3. Securing PLCs

PLCs are particularly sensitive and vulnerable equipment. In addition to the general measures for workstations that may apply, the basic measures to be implemented on this equipment include:

– protection of logical access to PLCs by a password, which must not be the default password;

– protection of physical access to the PLCs (locked cabinet) and, if necessary, detection of opening by sending an alarm;

– the deactivation of remote configuration and/or remote programming modes when this possibility exists;

– the configuration of read-only access for first-level maintenance interventions, where this functionality exists;

– protection of access to source code and embedded code in CPUs;

– disabling unnecessary services (web, for example);

– disabling unused logical ports;

– the restriction of the IP addresses that can be connected.

11.12.4. *Securing IIoT equipment*

In addition to measures similar to conventional computer equipment, it is recommended to follow good practices such as those of the Industrial Internet Consortium (IIC) (Chapter 6) (Hanna *et al.* 2018), including the use of devices with a secure element (Chapter 10, section 10.7) (Hauet n.d.b).

In addition, it should be noted that new approaches based on the blockchain are promising in terms of solutions for securing IIoT networks (Appendix 2).

11.12.5. *Securing network equipment*

Network equipment including routers, gateways, switches and firewalls, industrial or not, are equipment with a computer and an operating system inside. It is therefore necessary to monitor vulnerabilities that involve them, update firmware and implement a password management policy for these devices (change of default password, periodic change and according to staff movements).

IoT gateways are special devices that provide communication between internal networks of IoT devices and the Internet. In particular, they are responsible for encryption. They must be configured to use TLS (Transport Layer Security, appendix 1.10) or another type of VPN such as IPsec and to use certificate-based authentication, both for internal and external communication.

The Cloud platform must be configured with the appropriate controls and supplied with the correct certificates.

11.12.6. *Antivirus*

Implementation of antivirus software is a basic security measure. Remember that this is a necessary measure, but is not sufficient alone because an antivirus does not necessarily detect all viruses. In addition, antivirus software is limited to certain operating systems (Windows or Mac OS) and does not exist for PLCs and other OT equipment.

To be effective, antivirus protection must be up to date. The update in the OT zone must be carried out without compromising the protection provided by the architecture and perimeter protection. Installation on all workstations in the OT zone is therefore not necessarily to be preferred. An analysis must be carried out to identify the stations to be equipped and those that will not be equipped and will be made safe by a hardening of the configuration, as they are in areas not directly connected, or are subject to particular constraints related to operational safety (availability).

The deployment and update policy must include:

– deployment on all mobile stations, engineering stations, remote maintenance stations and SCADA stations (if it does not have compatibility problems);

– antivirus scanning at the firewall level;

– an automatic update of non-critical items;

– a manual update of sensitive items with a fixed frequency;

– special care for the monitoring of demilitarized zone (DMZ) area stations (which can be attack relays);

– the deployment of an antivirus server in the DMZ area.

11.13. Data security and configuration

Data and configurations of OT equipment (programs and parameters) must be managed specifically. It is necessary to set up a procedure to manage the integrity of this data and compare the current running versions and those saved in the development stations.

In addition, when a new version is put into service, it is necessary to check that the differences between the previous version and the new version correspond to the changes made.

Commercial configuration management software can make this task easier.

In addition, all important data (firmware, standard software, SCADA software packages, PLC programs, configurations, etc.) should be saved and exchanged with an authenticity verification mechanism (signature).

11.14. Securing logical accesses

The security of logical access, based on user access control, is based on more or less sophisticated technical means and a procedure for using these means by defining the roles and rights of each user. The security measures to be implemented are as follows:

– access permissions should be managed on a role-based basis, incorporating the principle of least privilege and segregation of duties. Depending on the areas, an analysis must determine who can access them and who to what resources;

– identities and identifying information must be issued, managed, verified, revoked and audited for authorized devices, users and processes;

– remote access must be managed;

– users and equipment must be authenticated with one- or two-factor methods depending on the level of risk.

Older, or less powerful, hardware does not always have very advanced authentication capabilities, which can be a problem. In addition, it is often difficult to implement centralized authentication management, such as LDAP (Lightweight Directory Access Protocol, a protocol for querying directory services).

From a practical point of view, passwords should be chosen according to appropriate rules to provide a good level of security, and default passwords such as admin/admin should be changed.

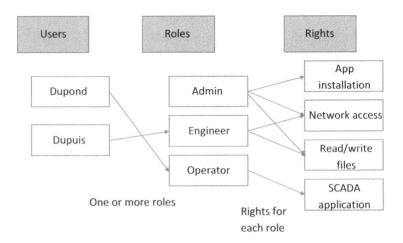

Figure 11.7. *Access control by role*

11.15. Securing supplier and service provider interactions

In all its life phases, the security of an ICS depends heavily on a number of external suppliers or service providers. Recent attacks as Stuxnet have demonstrated this.

To guarantee the desired level of security, special requirements apply to these stakeholders. These have been standardized (IEC 62443-2-4 standard) or are given in various guides (ANSSI 2013a).

Several aspects can be identified:

– component procurement: off-the-shelf components (COTS) may contain vulnerabilities, whether intentional, such as backdoors, or not, simply because they are due to a software development process that does not take security into account. To prevent this risk, it is possible to use suppliers or products certified with CSPN (ANSSI n.d.), ISA Secure (IEC n.d.) and TUV (TÜV Certification n.d.) certifications for example;

– outsourcing and external service providers: these providers must at least be trained. For high-risk sites or OESs, providers must be qualified (qualified trusted providers).

11.16. Incident detection

11.16.1. *Logging and alerts*

To complement the protection measures, an incident detection system must be implemented. Most systems (workstations, servers, network equipment) have the ability to log different types of events. These possibilities must be activated.

The questions that arise are the types of events to be recorded and the storage period in order to limit the size of storage. These logs can be centralized with a SIEM.

11.16.2. *Intrusion detection system*

To monitor the network and workstations, it is possible to set up an IDS. These systems are presented in Chapter 10.

11.16.3. *Centralization of events (SIEM)*

To process events centrally, there are systems called SIEM, presented in Chapter 10. These can be coupled with SCADA supervision to generate alerts.

11.17. Security monitoring

11.17.1. *Updating mapping and documentation*

The system can be upgraded, especially with IIoT architectures or mobile devices. A procedure to keep the mapping up to date, possibly supplemented by a computer tool for detecting new equipment, should be put in place.

In parallel, the documentation (section 11.4.2) must be kept up to date.

11.17.2. *Security patch management*

New vulnerabilities in hardware and software are regularly discovered and published by Computer Emergency Response Teams (CERTs) (Chapter 5).

The updating of security patches must be done within the framework of a controlled process, with the implementation of an adapted organization. Compared to IT systems, automatic updating of software patches is not always possible, on the one hand, because equipment does not always have this possibility, and, on the other hand, because the risk of disrupting production is high. The patch management policy must be adapted to the risks and constraints. It can be adapted to different types of equipment, with, for example:

– a systematic update of the workstations;

– a periodic update or during maintenance for PLCs.

For workstation updates, it may be preferable to use the DMZ.

The monitoring activity is based on mapping, and it will be all the more effective if the mapping is exhaustive.

11.17.3. *Audit of the facility*

An audit of the installation can be carried out regularly using the approach described in Chapter 5, and regular penetration tests during maintenance periods can also be considered to avoid disrupting normal operation.

11.18. Incident handling

Incidents detected by systems implemented in the previous phase should be appropriately addressed. This is defined by the implementation of an organization and procedures that make it possible to determine:

– what to do when an incident is detected;

– who must be alerted according to the level of importance of the incident;

– which measures should be applied in the response phase.

The response to incidents in the ICS case must take into account the physical aspect of the process: it is not possible to stop it in any situation and a mechanism must be provided to bring it back to a safe situation, or to provide for operation in degraded mode. The answer must also take into account the fact that the system operates in real time, and be fast enough.

In the event of an incident, responders must be trained in the specific scenarios of the ICS, as normal methods of recovering computer systems may not be applicable to an ICS.

An escalation management process can be defined to manage incidents at the right level of responsibility.

The incident management process must also include a post-incident analysis phase. This feedback allows for continuous improvement in risk management.

11.19. Recovery

If certain feared events occur, damage may occur to the ICS and the installation. It is therefore important to be prepared to restart the activity as soon as possible. Two levels of damage can be considered:

– the first one concerns programs and data, and a recovery allows to restart;

– the second concerns physical damage and, in this case, more extensive rehabilitation is required.

11.19.1. *Backup*

A backup policy must be defined. It includes defining the scope of the elements to be backed up, which includes:

– images of servers and workstations;

– configuration databases (users, alarm thresholds, etc.);

– PLC programs (sources) and data;

– the control parameters and firmware of intelligent sensors and actuators;

– SCADA historian database;

– PLC firmware, configuration files and network equipment firmware (switch, VPN, router, firewall, etc.).

In a second step, it is necessary to define the periodicity of the backup and the storage duration. Regulatory requirements may need to be considered.

Finally, it is necessary to define the means to perform the backup, which can be automated, but not always for some OT equipment settings. A traceability of the modifications is therefore to be expected.

The last step is to provide for regular backup testing procedures.

11.19.2. *Business continuity plan*

A business continuity plan is designed to ensure the company's survival in the event of a disaster. It must therefore integrate industrial information systems. It includes defining the Disaster Recovery Plan, which identifies the means and procedures necessary to return to a nominal situation as quickly as possible in the event of significant damage. It also describes how to rebuild the system following a cyber-attack, fire, flood or data loss.

11.20. Cybersecurity and lifecycle

The measures presented are implemented in the operation and maintenance phase of the installation. However, it is important to anticipate this cybersecurity issue from the early stages in the design phase and integration phase. A process of transfer of responsibility for cybersecurity and useful information about it must be provided during the commissioning. Finally, cybersecurity must also be managed during the deconstruction phase, one of the well-known vulnerabilities being the disposal of devices containing confidential data.

Appendix 1

Cryptography Basics

A1.1. Introduction

Cryptography is one of the basic techniques for securing information systems. Content is encrypted using a method or algorithm that transforms a message to make it unintelligible. This technique is old and has existed since ancient times. Its weak point was the need to transmit the encryption and decryption method to all parties. In 1976, Whitfield Diffie and Martin Hellman (Diffie *et al.* 1976), of Stanford University, proposed an entirely new encryption principle: public key or asymmetric cryptography. This technique allows you to distribute a key to encrypt a message, but this key does not allow you to decrypt it. This principle has been incorporated into the RSA system (Rivest–Shamir–Adleman, names of inventors).

A1.1.1. *Definitions of the terms*

Encryption is about transforming a message in such a way that it becomes incomprehensible. Only authorized correspondents will be able to decipher it.

Data encryption consists of applying a function F configurable by a key k to a message M, so that the message cannot be decrypted.

Decryption consists of transforming a previously encrypted message to reconstitute the original message. The objective is that only authorized correspondents can carry out this action.

Unauthorized correspondents aims at reconstituting the original message by trying to "break" the cryptographic code or algorithm.

Signing a document consists of creating an electronic signature unique to the document and its author. The signature thus links the original document and its author.

Verifying the signature is to ensure that the original document has not been modified and that its author is authenticated. If the signature is not valid, then the document should not be trusted.

A1.2. Symmetric cryptography

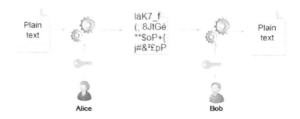

Figure A1.1. *Symmetric encryption*

The key used for encryption is the same as the one used for decryption. This key must be secret: only authorized persons must possess it, otherwise the confidentiality of the message is no longer guaranteed.

The disadvantages of this approach are the risk of loss or theft of the key, especially when it is communicated to the correspondent in the initial phase.

A1.3. Asymmetric cryptography

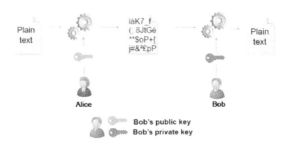

Figure A1.2. *Asymmetric encryption*

The key used for encryption is different from the one used for decryption. It is necessary to use two keys:

– a public key: as its name suggests, this key is public and can be given to everyone. It is used to encrypt messages;

– a private key: this key must be personal and known only to its owner. It must remain secret. It is used to decipher messages.

The principle is as follows: if Alice is to receive a message from Bob, but she does not trust the postman who could open her letter, she will first send Bob an open box with an encrypted lock, of which she alone has the code. Bob will put his message in the box and close it, before sending it to Alice. The postman will not be able to open the box, since only Alice has the key. Thus, a public key cryptography system is actually based on *two keys*:

– a public key, which can be freely distributed, is the open box;

– a secret key, known only to the recipient, is the box code.

This is why we talk about asymmetric encryption.

From a mathematical point of view, we have a function P on integers, which has an inverse S. It is assumed that such a pair (P, S) can be built, but that, knowing only P, it is impossible (or at least very difficult) to find S. In other words, it is necessary to mathematically determine functions that are difficult to reverse, or "one-way". Modular exponentiation is such a function and is used in the Diffie-Hellman method (Diffie *et al.* 1976).

Asymmetric encryption is used in conjunction with symmetric encryption, often in an initial phase to exchange keys for symmetric encryption.

Symmetric encryption	Asymmetric encryption
Fast and adapted to real-time traffic	More computationally expensive and used mainly for the initial key exchange phase before moving to symmetric encryption
Short keys (currently 256 bits)	Longer keys (currently 2,048 bits minimum)
Example of an algorithm: AES (Daemen and Rijmen 1999)	Example of an algorithm: RSA (Rivest *et al.* 1978)

Table A1.1. *The different aspects covered by the standard*

A1.4. MD5 and SHA cryptographic footprints (hash codes)

MD5 and SHA (SHA-1, SHA-2, SHA 256, etc.) are one-way cryptographic hash functions. The result of applying this function to a data set of any length (e.g. a file) is a much smaller set of fixed length, characteristic of the initial dataset, called a checksum, a condensate, a cryptographic fingerprint, a hash code or a fingerprint.

Algorithm	Condensate length
CRC32	32 bits (no longer used)
MD5	128 bits
SHA-1	160 bits
SHA-2: family of algorithms (SHA-256, SHA-224, SHA-384, or SHA-512)	From 256 to 512 bits

Table A1.2. *Some hash algorithms*

For example, with MD5, we get:

– "This is a test": 11b35a0201513381dcdd130831f702d0;

– for the first paragraph of this section: 730c7a6d6bad97c975f990aba8d329e3.

The objective is to have a unique and irreversible footprint. Since initial algorithms such as SHA-1 have produced collisions (non-uniqueness of the result, i.e. the same fingerprint can be obtained for two different data sets), the most recent algorithms provide longer numbers, e.g. SHA-256 provides 256 bits

A1.5. Electronic signature

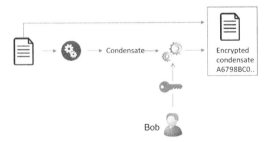

Figure A1.3. *Signing a document*

The purpose of a signature is to ensure that the data are not modified, and to verify the identity of its author. If the signature is not valid, it indicates that the author is not the one identified, or that the data received are not those signed by the author.

If confidentiality is to be ensured, it is necessary to encrypt the document.

The principle of electronic signature is as follows:

– the sender of a message generates a condensate from the message he wishes to sign using a fingerprint generation algorithm;

– the sender uses his private key to encrypt the previous condensate and to produce an electronic signature;

– the sender sends (or stores) the message and the electronic signature.

To verify the signature, the correspondent calculates the condensate from the plaintext message and compares it with the decrypted signature, with the public key of the correspondent.

A1.6. Certificates

A certificate allows a public key to be associated with an entity in a guaranteed way. This is useful to ensure that a caller's public key is his or her own or that a website is not fraudulent.

The X509 standard provides a standard certificate format.

A certificate is an electronic file that includes, among other things:

– the public key of a person or entity or domain name;

– the description of this person or entity: name, first name, domain name, etc.;

– signature by a trusted third party, responsible for ensuring that the owner of the public key has been verified and, consequently, the authenticity of the public key as that of the owner. The signature refers to the identity of the holder and the public key in order to ensure the integrity of the whole other information such as key usage, validity dates and revocation information.

The trusted third party is a certification authority (CA) in charge of:

– verifying the identity of the person requesting to create the certificate;

– creating the certificate after verification, then signing it (with the private key from the CA);

– maintaining a list of certificates that have been revoked (for example, if the key has been compromised).

There are trusted third parties that are known to browsers by default. Certification authorities can be organized hierarchically.

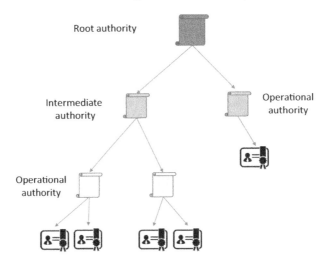

Figure A1.4. *Hierarchy of authorities*

A1.7. Public Key Infrastructure architecture

A Public Key Infrastructure (PKI) is a public key management system that manages important lists of public keys and ensures their reliability, usually for entities in a network. A PKI allows public keys to be linked to identities (such as user or organization names). A PKI provides guarantees that a public key obtained through it can be trusted *a priori*, but it is not a CA.

An architecture can be based on certification authorities, but can also use other mechanisms such as a blockchain.

A PKI manages the security certificate lifecycle to associate an entity with a public key. It responds to requests for certificate verification and manages the revocation of certificates.

A1.8. Public Key Cryptography Standards

Public Key Cryptography Standards (PKCS) are a set of specifications designed by RSA laboratories in California.

A1.9. Secure SHell

Secure SHell (SSH) is the replacement of various shell-based protocols such as Telnet, as well as file transfer or copy protocols such as File Transfer Protocol (FTP) and Remote Copy (RCP). SSH can be used to create a tunnel for other applications.

SSH uses public key cryptography to prove the authenticity of the remote user. SSH can generate a key pair. It uses a fingerprint that is a snapshot of an individual host's real public key (for example, the RSA public part). Fingerprints are usually 128 bits long. That's what the user can use to verify that a public key is that of an individual or a host. SSH maintains a list of trusted hosts. Communication of real data is secured using symmetric cryptography such as AES or 3DES, IDEA.

A1.10. TLS/SSL

TLS/SSL (Transport Layer Security/Secure Sockets Layer) ensures the security of the transport layer. TLS/SSL is not a single application like SSH, but provides security through its implementation in applications. SSL was designed by Netscape to secure the use of HTTP. TLS is the latest version of SSL technology. The procedure for a connection is as follows:

– the client's browser sends a request to the server to set up a connection secured by TLS;

– the server responds by sending its certificate;

– three scenarios may occur:

 - the certificate signature can be verified with a certificate from a CA integrated by default in the browser. The client can check the certificate with the CA;

- the signature can be verified with the server's public key: the certificate is self-signed. In this case, there is a possibility that the site is fraudulent. The user is informed and validates or does not validate continuation of the connection;

- the signature cannot be verified, the connection cannot continue;

– if the certificate is validated, the client generates an encryption key that he/she encrypts using the public key contained in the server certificate, then transmits this session key to the server to continue the exchange.

A1.11. IPsec

The IPsec protocol provides security on the IP packet layer; it is integrated at a lower level than TLS/SSL in the Transmission Control Protocol (TCP)/Internet protocol (IP) stack (Chapter 2). IPsec is a network level protocol embedded in servers and/or clients, for example, in a router, dedicated VPN hub, firewall or operating system core.

It is important to note that there is no interoperability between SSH, TLS/SSL and IPsec, as they all operate at different levels in the TCP model and are designed for different uses.

A communication between two hosts protected by IPsec is likely to operate in two different modes: transport mode and tunnel mode. In the transport mode, only the transferred data (the payload part of the IP packet) is encrypted and/or authenticated. The rest of the IP packet is unchanged and, as a result, the routing of the packets is not changed.

In tunnel mode, the entire IP packet is encrypted and/or authenticated. The packet is then encapsulated in a new IP packet with a new IP header. The second mode allows IP datagrams to be encapsulated in other IP datagrams, the content of which is protected. The major advantage of this second mode is that it makes it possible to set up security gateways that handle the entire IPsec aspect of a communication and transmit the purified datagrams from their IPsec part to their real recipient.

Appendix 2

Blockchain and IIoT Security

A2.1. Blockchain principle

The first blockchain was developed for the Bitcoin cryptocurrency. It appears that the concept is quite broad and the applications of blockchain are very numerous.

A blockchain operates using Distributed Ledger Technology (DLT). It is a database system with specific features:

– it stores transactions in a distributed way by replicating them on the nodes of a peer-to-peer network (P2P);

– it operates in a decentralized manner, i.e. without a central authority or control body;

– it is secure and stores transactions immutably in blocks. Each transaction is verified and recorded using cryptographic techniques to create a history in the form of a chain of blocks.

Records in which transactions were recorded existed before the birth of computer technology in the paper form, particularly in accounting (ledger is the term used in accounting). A central authority was in charge of keeping it and checking the validity of the entries.

With DLT technologies, registry management offers additional capabilities, since it operates without central authority and in a distributed manner.

The blockchain uses the ECC cryptography algorithm (cryptography on elliptic curves) and the SHA-256 hash to provide solid cryptographic evidence for data authentication and integrity.

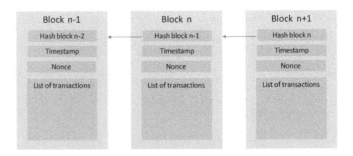

Figure A2.1. *Blockchain*

Each block contains a list of transactions and a hash from the previous block. The blockchain therefore has a history of all transactions, and constitutes an unforgeable shared register available to all, without central authority.

In the Bitcoin chain, a block is validated by determining a parameter (called nonce) that makes it possible to obtain a hash starting with a certain number of zeros. Miners compete to solve this problem. The solution found is called Proof of Work. The number of zeros defines the degree of difficulty of the validation. For the Bitcoin infrastructure, it varies so that for the average mining time is around 10 min.

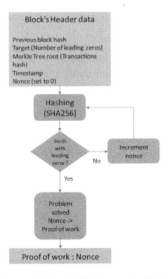

Figure A2.2. *Validation of a block*

The concept initially proposed by the Bitcoin blockchain has been generalized. Different approaches exist, and they are characterized by:

– the mechanism to guarantee trust, called the consensus mechanism, which makes it possible to validate a block and guarantee its integrity. There are mechanisms other than proof of work such as proof of stake;

– the mechanism for propagating and synchronizing blocks between network nodes. In addition to the Bitcoin method, there are other solutions such as propagation based on headers, or unsolicited sending.

The choice of the approach depends on the degree of opening of the blockchain. Indeed, there are public blockchains, blockchains with controlled access and private blockchains. Depending on the type, the level of trust of the nodes is different and, therefore, the consensus mechanism is adapted. Whenever possible, the type that is the least time consuming to calculate is chosen.

A2.2. Smartcontracts

The notion of smartcontracts was proposed by the blockchain Ethereum. It was launched in July 2015 and since then a large number of similar platforms have emerged, such as HyperLedger.

Unlike the Bitcoin blockchain, which is mainly used for digital currency transactions, the Ethereum blockchain has the ability to store records and, more importantly, to execute smart contracts. The term "intelligent contracts" was invented by Nick Szabo in 1994.

An intelligent contract is an IT protocol designed to facilitate, verify or digitally apply the negotiation or execution of a contract. Smart contracts allow credible transactions to be carried out without third parties. These transactions are traceable and irreversible.

More concretely, a smart contract is a piece of code that resides on a block string and is identified by a unique address. An intelligent contract includes a set of executable functions and state variables.

Any user of the blockchain network can trigger the contract functions by sending transactions to the contract. Transactions include input parameters required by the contract functions. When executing a function, the contract status variables change according to the logic implemented in the function. Contracts can be written in different high-level languages (such as Solidity close to Javascript). The Ethereum blockchain allows the use of Ethereum Virtual Machines, which are the mining nodes and which are capable of ensuring a reliable and inviolable execution of these programs or contracts.

The contract code is executed on each node participating in the network as part of the verification of new (mining) blocks.

A2.3. The contributions of the blockchain for the Industrial Internet of Things

A blockchain is a distributed and secure database system that does not use trusted third parties and capable of performing secure actions. This type of infrastructure could therefore provide solutions to the problems of securing Industrial Internet of Things (IIoTs), and much work is currently being done to explore its possibilities.

A2.3.1. *PKI Infrastructure*

Since blockchain technology provides a distributed and unalterable information registry, it appears to be potentially appropriate for public key storage and management. This could allow for better availability and robustness compared to using a centralized server.

A2.3.2. *Object identification*

The blockchain has an address space of 160 bits, more than the IP6 space, which is 128 bits. It can therefore provide an address for 1.46×10^{48} pieces of equipment (Globally Unique Identifier [GUID]). An identification using the blockchain would be decentralized and would no longer require a centralized system for address distribution.

For management of the ownership of assets (equipment), the *blockchain* can also provide a reliable solution. It can be used to register and give identity to connected IoT devices, with a set of complex attributes and relationships that can be downloaded and stored on the distributed blockchain registry. Among these attributes, it is possible to include successive owners.

A2.3.3. *Communications security*

With a blockchain, the management and distribution of keys is completely eliminated, as each IoT device will have its own GUID and asymmetric key pair when it connects to the blockchain network. This can also significantly simplify other security protocols, since it is not necessary to manage and exchange PKI certificates during the initialization phase of the transaction.

A2.3.4. *Firmware update*

By using smartcontracts, which are secure because they are in the blockchain, and activated by a transaction, the secure distribution of updates is greatly simplified.

A2.3.5. *Automatic execution of actions*

With a smartcontract, a self-executable code that can be integrated into each IoT chip could determine what action to take when a condition is met. These actions would only be performed if an incoming transaction was authenticated.

For example, an intelligent sensor could initiate automatic actions, or a supply-chain system could be coupled with invoicing.

A2.4. The limitations of blockchain

Blockchain does not solve all problems and this promising technology is still in its infancy. Currently, a number of limitations exist. This technology is expensive in terms of network bandwidth and computing time (and therefore energy). It can have a relatively long response time compared to a centralized architecture.

In addition, blockchain has vulnerabilities (Li et al., 2017): for example, it is enough to convince 51% of the nodes to make a fact a truth, which can be a problem on small networks.

Appendix 3

NIST SP 800-82 Security Measures

This appendix presents the security measures proposed in the NIST SP800-82r4 guide. More details are given in this guide (Stouffer *et al.* 2015), with guidance on the implementation of these measures for industrial control system (ICS). The measurements are aligned with those proposed in SP 800-53 (NIST 2014).

ACCESS CONTROL – AC				
AC-1	**Access Control Policy and Procedures**	x	x	x
AC-2	**Account Management**	x	x	x
AC-2 (1)	*ACCOUNT MANAGEMENT	AUTOMATED SYSTEM ACCOUNT MANAGEMENT*		x	x
AC-2 (2)	*ACCOUNT MANAGEMENT	REMOVAL OF TEMPORARY/EMERGENCY ACCOUNTS*		x	x
AC-2 (3)	*ACCOUNT MANAGEMENT	DISABLE INACTIVE ACCOUNTS*		x	x
AC-2 (4)	*ACCOUNT MANAGEMENT	AUTOMATED AUDIT ACTIONS*		x	x
AC-2 (5)	*ACCOUNT MANAGEMENT	INACTIVITY LOGOUT/TYPICAL USAGE MONITORING*			x
AC-2 (11)	*ACCOUNT MANAGEMENT	USAGE CONDITIONS*			x
AC-2 (12)	*ACCOUNT MANAGEMENT	ACCOUNT MONITORING/ATYPICAL USAGE*			x

AC-2 (13)	*ACCOUNT MANAGEMENT | ACCOUNT REVIEWS*			x
AC-3	**Access Enforcement**	x	x	x
AC-4	**Information Flow Enforcement**		x	x
AC-5	**Separation of Duties**		x	x
AC-6	**Least Privilege**		x	x
AC-6 (1)	*LEAST PRIVILEGE | AUTHORIZE ACCESS TO SECURITY FUNCTIONS*		x	x
AC-6 (2)	*LEAST PRIVILEGE | NON-PRIVILEGED ACCESS FOR NONSECURITY FUNCTIONS*		x	x
AC-6 (3)	*LEAST PRIVILEGE | NETWORK ACCESS TO PRIVILEGED COMMANDS*			x
AC-6 (5)	*LEAST PRIVILEGE | PRIVILEGED ACCOUNTS*		x	x
AC-6 (9)	*LEAST PRIVILEGE | AUDITING USE OF PRIVILEGED FUNCTIONS*		x	x
AC-6 (10)	*LEAST PRIVILEGE | PROHIBIT NON-PRIVILEGED USERS FROM EXECUTING PRIVILEGED FUNCTIONS*		x	x
AC-7	**Unsuccessful Login Attempts**	x	x	x
AC-8	**System Use Notification**	x	x	x
AC-10	Concurrent Session Control			x
AC-11	**Session Lock**		x	x
AC-11 (1)	*SESSION LOCK | PATTERN-HIDING DISPLAYS*		x	x
AC-12	**Session Termination**		x	x
AC-14	**Permitted Actions without Identification or Authentication**	x	x	x
AC-17	**Remote Access**	x	x	x
AC-17 (1)	*REMOTE ACCESS | AUTOMATED MONITORING/CONTROL*		x	x

AC-17 (2)	REMOTE ACCESS | PROTECTION OF CONFIDENTIALITY/INTEGRITY USING ENCRYPTION		x	x
AC-17 (3)	REMOTE ACCESS | MANAGED ACCESS CONTROL POINTS		x	x
AC-17 (4)	REMOTE ACCESS | PRIVILEGED COMMANDS/ACCESS		x	x
AC-18	Wireless Access	x	x	x
AC-18 (1)	WIRELESS ACCESS | AUTHENTICATION AND ENCRYPTION		x	x
AC-18 (4)	WIRELESS ACCESS | RESTRICT CONFIGURATIONS BY USERS			x
AC-18 (5)	WIRELESS ACCESS | CONFINE WIRELESS COMMUNICATIONS			x
AC-19	Access Control for Mobile Devices	x	x	x
AC-19 (5)	ACCESS CONTROL FOR MOBILE DEVICES | FULL DEVICE/CONTAINER-BASED ENCRYPTION		x	x
AC-20	Use of External Information Systems	x	x	x
AC-20 (1)	USE OF EXTERNAL INFORMATION SYSTEMS | LIMITS ON AUTHORIZED USE		x	x
AC-20 (2)	USE OF EXTERNAL INFORMATION SYSTEMS | PORTABLE STORAGE MEDIA		x	x
AC-21	Collaboration and Information Sharing	+	x	x
AC-22	Publicly Accessible Content	x	x	x
AWARENESS AND TRAINING – AT				
AT-1	Security Awareness and Training Policy and Procedures	x	x	x
AT-2	Security Awareness	x	x	x
AT-2 (2)	SECURITY AWARENESS | INSIDER THREAT		x	x
AT-3	Role-Based Security Training	x	x	x
AT-4	Security Training Records	x	x	x

	AUDITING AND ACCOUNTABILITY – AU			
AU-1	Audit and Accountability Policy and Procedures	x	x	x
AU-2	Auditable Events	x	x	x
AU-2 (3)	*AUDITABLE EVENTS	REVIEWS AND UPDATES*		x	x
AU-3	Content of Audit Records	x	x	x
AU-3 (1)	*CONTENT OF AUDIT RECORDS	ADDITIONAL AUDIT INFORMATION*		x	x
AU-3 (2)	*CONTENT OF AUDIT RECORDS	CENTRALIZED MANAGEMENT OF PLANNED AUDIT RECORD CONTENT*			x
AU-4	Audit Storage Capacity	x	x	x
AU-4 (1)	*AUDIT STORAGE CAPACITY	TRANSFER TO ALTERNATE STORAGE*	+	+	+
AU-5	Response to Audit Processing Failures	x	x	x
AU-5 (1)	*RESPONSE TO AUDIT PROCESSING FAILURES	AUDIT STORAGE CAPACITY*			x
AU-5 (2)	*RESPONSE TO AUDIT PROCESSING FAILURES	REAL-TIME ALERTS*			x
AU-6	Audit Review, Analysis and Reporting	x	x	x
AU-6 (1)	*AUDIT REVIEW, ANALYSIS AND REPORTING	PROCESS INTEGRATION*		x	x
AU-6 (3)	*AUDIT REVIEW, ANALYSIS AND REPORTING	CORRELATE AUDIT REPOSITORIES*		x	x
AU-6 (5)	*AUDIT REVIEW, ANALYSIS AND REPORTING	INTEGRATION/ SCANNING AND MONITORING CAPABILITIES*			x
AU-6 (6)	*AUDIT REVIEW, ANALYSIS AND REPORTING	CORRELATION WITH PHYSICAL MONITORING*			x
AU-7	Audit Reduction and Report Generation		x	x
AU-7 (1)	*AUDIT REDUCTION AND REPORT GENERATION	AUTOMATIC PROCESSING*		x	x
AU-8	Time Stamps	x	x	x

AU-8 (1)	TIME STAMPS | SYNCHRONIZATION WITH AUTHORITATIVE TIME SOURCE		x	x
AU-9	**Protection of Audit Information**	x	x	x
AU-9 (2)	PROTECTION OF AUDIT INFORMATION | AUDIT BACKUP ON SEPARATE PHYSICAL SYSTEMS/COMPONENTS			x
AU-9 (3)	PROTECTION OF AUDIT INFORMATION | CRYPTOGRAPHIC PROTECTION			x
AU-9 (4)	PROTECTION OF AUDIT INFORMATION | ACCESS BY SUBSET OF PRIVILEGED USERS		x	x
AU-10	Non-repudiation			x
AU-11	**Audit Record Retention**	x	x	x
AU-12	**Audit Generation**	x	x	x
AU-12 (1)	AUDIT GENERATION | SYSTEM-WIDE/TIME-CORRELATED AUDIT TRAIL			x
AU-12 (3)	AUDIT GENERATION | CHANGES BY AUTHORIZED INDIVIDUALS			x
SECURITY ASSESSMENT AND AUTHORIZATION – CA				
CA-1	**Security Assessment and Authorization Policy and Procedures**	x	x	x
CA-2	**Security Assessments**	x	x	x
CA-2 (1)	SECURITY ASSESSMENTS | INDEPENDENT ASSESSORS		x	x
CA-2 (2)	SECURITY ASSESSMENTS | TYPES OF ASSESSMENTS			x
CA-3	**Information System Connections**	x	x	x
CA-3 (5)	SYSTEM INTERCONNECTIONS | RESTRICTIONS ON EXTERNAL SYSTEM CONNECTIONS		x	x
CA-5	**Plan of Action and Milestones**	x	x	x
CA-6	**Security Authorization**	x	x	x
CA-7	**Continuous Monitoring**	x	x	x
CA-7 (1)	CONTINUOUS MONITORING | INDEPENDENT ASSESSMENT		x	x
CA-8	**Penetration Testing**			x
CA-9	**Internal System Connections**	x	x	x

	CONFIGURATION MANAGEMENT – CM			
CM-1	Configuration Management Policy and Procedures	x	x	x
CM-2	Baseline Configuration	x	x	x
CM-2 (1)	BASELINE CONFIGURATION | REVIEWS AND UPDATES		x	x
CM-2 (2)	BASELINE CONFIGURATION | AUTOMATION SUPPORT FOR ACCURACY / CURRENCY			x
CM-2 (3)	BASELINE CONFIGURATION | RETENTION OF PREVIOUS CONFIGURATIONS		x	x
CM-2 (7)	BASELINE CONFIGURATION | CONFIGURE SYSTEMS, COMPONENTS, OR DEVICES FOR HIGH-RISK AREAS		x	x
CM-3	Configuration Change Control		x	x
CM-3 (1)	CONFIGURATION CHANGE CONTROL | AUTOMATED DOCUMENT / NOTIFICATION/PROHIBITION OF CHANGES			x
CM-3 (2)	CONFIGURATION CHANGE CONTROL | TEST/VALIDATE/DOCUMENT CHANGES		x	x
CM-4	Security Impact Analysis	x	x	x
CM-4 (1)	SECURITY IMPACT ANALYSIS | SEPARATE TEST ENVIRONMENTS			x
CM-5	Access Restrictions for Change		x	x
CM-5 (1)	ACCESS RESTRICTIONS FOR CHANGE | AUTOMATED ACCESS ENFORCEMENT/AUDITING			x
CM-5 (2)	ACCESS RESTRICTIONS FOR CHANGE | AUDIT SYSTEM CHANGES			x
CM-5 (3)	ACCESS RESTRICTIONS FOR CHANGE | SIGNED COMPONENTS			x
CM-6	Configuration Settings	x	x	x
CM-6 (1)	CONFIGURATION SETTINGS | AUTOMATED CENTRAL MANAGEMENT / APPLICATION/VERIFICATION			x

CM-6 (2)	CONFIGURATION SETTINGS | RESPOND TO UNAUTHORIZED CHANGES			x
CM-7	**Least Functionality**	x	x	x
CM-7 (1)	LEAST FUNCTIONALITY | PERIODIC REVIEW	+	x	x
CM-7 (2)	LEAST FUNCTIONALITY | PREVENT PROGRAM EXECUTION		x	x
CM-7 (4)	LEAST FUNCTIONALITY | UNAUTHORIZED SOFTWARE		–	
CM-7 (5)	LEAST FUNCTIONALITY | AUTHORIZED SOFTWARE		+	x
CM-8	**Information System Component Inventory**	x	x	x
CM-8 (1)	INFORMATION SYSTEM COMPONENT INVENTORY | UPDATES DURING INSTALLATIONS/REMOVALS		x	x
CM-8 (2)	INFORMATION SYSTEM COMPONENT INVENTORY | AUTOMATED MAINTENANCE			x
CM-8 (3)	INFORMATION SYSTEM COMPONENT INVENTORY | AUTOMATED UNAUTHORIZED COMPONENT DETECTION		x	x
CM-8 (4)	INFORMATION SYSTEM COMPONENT INVENTORY | PROPERTY ACCOUNTABILITY INFORMATION			x
CM-8 (5)	INFORMATION SYSTEM COMPONENT INVENTORY | ALL COMPONENTS WITHIN AUTHORIZATION BOUNDARY		x	x
CM-9	**Configuration Management Plan**		x	x
CM-10	**Software Usage Restrictions**	x	x	x
CM-11	**User-Installed Software**	x	x	x
CONTINGENCY PLANNING – CP				
CP-1	**Contingency Planning Policy and Procedures**	x	x	x
CP-2	**Contingency Plan**	x	x	x
CP-2 (1)	CONTINGENCY PLAN | COORDINATE WITH RELATED PLANS		x	x
CP-2 (2)	CONTINGENCY PLAN | CAPACITY PLANNING			x

ID	Name			
CP-2 (3)	CONTINGENCY PLAN	RESUME ESSENTIAL MISSIONS/BUSINESS FUNCTIONS		x	x
CP-2 (4)	CONTINGENCY PLAN	RESUME ALL MISSIONS/BUSINESS FUNCTIONS			x
CP-2 (5)	CONTINGENCY PLAN	CONTINUE ESSENTIAL MISSIONS/BUSINESS FUNCTIONS			x
CP-2 (8)	CONTINGENCY PLAN	IDENTIFY CRITICAL ASSETS		x	x
CP-3	Contingency Training	x	x	x
CP-3 (1)	CONTINGENCY TRAINING	SIMULATED EVENTS			x
CP-4	Contingency Plan Testing	x	x	x
CP-4 (1)	CONTINGENCY PLAN TESTING	COORDINATE WITH RELATED PLANS		x	x
CP-4 (2)	CONTINGENCY PLAN TESTING	ALTERNATE PROCESSING SITE			x
CP-6	Alternate Storage Site		x	x
CP-6 (1)	ALTERNATE STORAGE SITE	SEPARATION FROM PRIMARY SITE		x	x
CP-6 (2)	ALTERNATE STORAGE SITE	RECOVERY TIME/POINT OBJECTIVES			x
CP-6 (3)	ALTERNATE STORAGE SITE	ACCESSIBILITY		x	x
CP-7	Alternate Processing Site		x	x
CP-7 (1)	ALTERNATE PROCESSING SITE	SEPARATION FROM PRIMARY SITE		x	x
CP-7 (2)	ALTERNATE PROCESSING SITE	ACCESSIBILITY		x	x
CP-7 (3)	ALTERNATE PROCESSING SITE	PRIORITY OF SERVICE		x	x
CP-7 (4)	ALTERNATE PROCESSING SITE	CONFIGURATION FOR USE			x
CP-8	Telecommunications Services		x	x
CP-8 (1)	TELECOMMUNICATIONS SERVICES	PRIORITY OF SERVICE PROVISIONS		x	x

ID	Control			
CP-8 (2)	TELECOMMUNICATIONS SERVICES | SINGLE POINTS OF FAILURE		X	X
CP-8 (3)	TELECOMMUNICATIONS SERVICES | SEPARATION OF PRIMARY / ALTERNATE PROVIDERS			X
CP-8 (4)	TELECOMMUNICATIONS SERVICES | PROVIDER CONTINGENCY PLAN			X
CP-9	**Information System Backup**	X	X	X
CP-9 (1)	INFORMATION SYSTEM BACKUP | TESTING FOR RELIABILITY / INTEGRITY		X	X
CP-9 (2)	INFORMATION SYSTEM BACKUP | TEST RESTORATION USING SAMPLING			X
CP-9 (3)	INFORMATION SYSTEM BACKUP | SEPARATE STORAGE FOR CRITICAL INFORMATION			X
CP-9 (5)	INFORMATION SYSTEM BACKUP | TRANSFER TO ALTERNATE SITE			X
CP-10	**Information System Recovery and Reconstitution**	X	X	X
CP-10 (2)	INFORMATION SYSTEM RECOVERY AND RECONSTITUTION | TRANSACTION RECOVERY		X	X
CP-10 (4)	INFORMATION SYSTEM RECOVERY AND RECONSTITUTION | RESTORE WITHIN TIME PERIOD			X
CP-12	**Safe Mode**	+	+	+
IDENTIFICATION AND AUTHENTICATION – IA				
IA-1	**Security Identification and Authentication Policy and Procedures**	X	X	X
IA-2	**Identification and Authentication (Organizational Users)**	X	X	X
IA-2 (1)	IDENTIFICATION AND AUTHENTICATION | NETWORK ACCESS TO PRIVILEGED ACCOUNTS	X	X	X
IA-2 (2)	IDENTIFICATION AND AUTHENTICATION | NETWORK ACCESS TO NON- PRIVILEGED ACCOUNTS		X	X

IA-2 (3)	IDENTIFICATION AND AUTHENTICATION | LOCAL ACCESS TO PRIVILEGED ACCOUNTS		x	x
IA-2 (4)	IDENTIFICATION AND AUTHENTICATION | LOCAL ACCESS TO NON- PRIVILEGED ACCOUNTS			x
IA-2 (8)	IDENTIFICATION AND AUTHENTICATION | NETWORK ACCESS TO PRIVILEGED ACCOUNTS – REPLAY RESISTANT		x	x
IA-2 (9)	IDENTIFICATION AND AUTHENTICATION | NETWORK ACCESS TO NON- PRIVILEGED ACCOUNTS – REPLAY RESISTANT			x
IA-2 (11)	IDENTIFICATION AND AUTHENTICATION | REMOTE ACCESS - SEPARATE DEVICE		x	x
IA-2 (12)	IDENTIFICATION AND AUTHENTICATION | ACCEPTANCE OF PIV CREDENTIALS	x	x	x
IA-3	Device Identification and Authentication	+	x	x
IA-3 (1)	DEVICE IDENTIFICATION AND AUTHENTICATION | CRYPTOGRAPHIC BIDIRECTIONAL AUTHENTICATION		+	+
IA-3 (4)	DEVICE IDENTIFICATION AND AUTHENTICATION | DEVICE ATTESTATION		+	+
IA-4	Identifier Management	x	x	x
IA-5	Authenticator Management	x	x	x
IA-5 (1)	AUTHENTICATOR MANAGEMENT | PASSWORD-BASED AUTHENTICATION	x	x	x
IA-5 (2)	AUTHENTICATOR MANAGEMENT | PKI-BASED AUTHENTICATION		x	x
IA-5 (3)	AUTHENTICATOR MANAGEMENT | IN PERSON REGISTRATION		x	x
IA-5 (11)	AUTHENTICATOR MANAGEMENT | HARDWARE TOKEN-BASED AUTHENTICATION	x	x	x
IA-6	Authenticator Feedback	x	x	x
IA-7	Cryptographic Module Authentication	x	x	x
IA-8	Identification and Authentication (Non-Organizational Users)	x	x	x

IA-8 (1)	*IDENTIFICATION AND AUTHENTICATION (NON-ORGANIZATIONAL USERS) | ACCEPTANCE OF PIV CREDENTIALS FROM OTHER AGENCIES*	x	x	x
IA-8 (2)	*IDENTIFICATION AND AUTHENTICATION (NON-ORGANIZATIONAL USERS) | ACCEPTANCE OF THIRD-PARTY CREDENTIALS*	x	x	x
IA-8 (3)	*IDENTIFICATION AND AUTHENTICATION (NON-ORGANIZATIONAL USERS) | USE OF FICAM-APPROVED PRODUCTS*	x	x	x
IA-8 (4)	*IDENTIFICATION AND AUTHENTICATION (NON-ORGANIZATIONAL USERS) | USE OF FICAM-ISSUED PROFILES*	x	x	x
	INCIDENT RESPONSE – IR			
IR-1	**Incident Response Policy and Procedures**	x	x	x
IR-2	**Incident Response Training**	x	x	x
IR-2 (1)	*INCIDENT RESPONSE TRAINING | SIMULATED EVENTS*			x
IR-2 (2)	*INCIDENT RESPONSE TRAINING | AUTOMATED TRAINING ENVIRONMENTS*			x
IR-3	**Incident Response Testing**		x	x
IR-3 (2)	*INCIDENT RESPONSE TESTING | COORDINATION WITH RELATED PLANS*		x	x
IR-4	**Incident Handling**	x	x	x
IR-4 (1)	*INCIDENT HANDLING | AUTOMATED INCIDENT HANDLING PROCESSES*		x	x
IR-4 (4)	*INCIDENT HANDLING | INFORMATION CORRELATION*			x
IR-5	**Incident Monitoring**	x	x	x
IR-5 (1)	*INCIDENT MONITORING | AUTOMATED TRACKING/DATA COLLECTION/ANALYSIS*			x
IR-6	**Incident Reporting**	x	x	x
IR-6 (1)	*INCIDENT REPORTING | AUTOMATED REPORTING*		x	x
IR-7	**Incident Response Assistance**	x	x	x

IR-7 (1)	*INCIDENT RESPONSE ASSISTANCE | AUTOMATION SUPPORT FOR AVAILABILITY OF INFORMATION/SUPPORT*			x	x
IR-8	**Incident Response Plan**	x	x	x	
MAINTENANCE – MA					
MA-1	**Maintenance Policy and Procedures**	x	x	x	
MA-2	**Controlled Maintenance**	x	x	x	
MA-2 (2)	*CONTROLLED MAINTENANCE | AUTOMATED MAINTENANCE ACTIVITIES*			x	
MA-3	**Maintenance Tools**		x	x	
MA-3 (1)	*MAINTENANCE TOOLS | INSPECT TOOLS*		x	x	
MA-3 (2)	*MAINTENANCE TOOLS | INSPECT MEDIA*		x	x	
MA-3 (3)	*MAINTENANCE TOOLS | PREVENT UNAUTHORIZED REMOVAL*			x	
MA-4	**Non-Local Maintenance**	x	x	x	
MA-4 (2)	*NON-LOCAL MAINTENANCE | DOCUMENT NON-LOCAL MAINTENANCE*		x	x	
MA-4 (3)	*NON-LOCAL MAINTENANCE | COMPARABLE SECURITY/SANITIZATION*			x	
MA-5	**Maintenance Personnel**	x	x	x	
MA-5 (1)	*MAINTENANCE PERSONNEL | INDIVIDUALS WITHOUT APPROPRIATE ACCESS*			x	
MA-6	**Timely Maintenance**		x	x	
MEDIA PROTECTION –MP					
MP-1	**Media Protection Policy and Procedures**	x	x	x	
MP-2	**Media Access**	x	x	x	
MP-3	**Media Marking**		x	x	
MP-4	**Media Storage**		x	x	
MP-5	**Media Transport**		x	x	
MP-5 (4)	*MEDIA TRANSPORT | CRYPTOGRAPHIC PROTECTION*		x	x	
MP-6	**Media Sanitization**	x	x	x	

MP-6 (1)	MEDIA SANITIZATION | TRACKING/DOCUMENTING/ VERIFYING			x
MP-6 (2)	MEDIA SANITIZATION | EQUIPMENT TESTING			x
MP-6 (3)	MEDIA SANITIZATION | NON-DESTRUCTIVE TECHNIQUES			x
MP-7	**Media Use**	x	x	x
MP-7 (1)	MEDIA USE | ORGANIZATIONAL RESTRICTIONS		x	x
PHYSICAL AND ENVIRONMENTAL PROTECTION – PE				
PE-1	**Physical and Environmental Protection Policy and Procedures**	x	x	x
PE-2	**Physical Access Authorizations**	x	x	x
PE-3	**Physical Access Control**	x	x	x
PE-3 (1)	PHYSICAL ACCESS CONTROL | INFORMATION SYSTEM ACCESS			x
PE-4	**Access Control for Transmission Medium**		x	x
PE-5	**Access Control for Output Devices**		x	x
PE-6	**Monitoring Physical Access**	x	x	x
PE-6 (1)	MONITORING PHYSICAL ACCESS | INTRUSION ALARMS / SURVEILLANCE EQUIPMENT		x	x
PE-6 (4)	MONITORING PHYSICAL ACCESS | MONITORING PHYSICAL ACCESS TO INFORMATION SYSTEMS		+	x
PE-8	**Visitor Access Records**	x	x	x
PE-8 (1)	VISITOR ACCESS RECORDS | AUTOMATED RECORDS MAINTENANCE / REVIEW			x
PE-9	**Power Equipment and Cabling**		x	x
PE-9 (1)	POWER EQUIPMENT AND CABLING | REDUNDANT CABLING		+	+
PE-10	**Emergency Shutoff**		x	x
PE-11	**Emergency Power**	+	x	x
PE-11 (1)	EMERGENCY POWER | LONG-TERM ALTERNATE POWER SUPPLY -	+	+	x

PE-11 (2)	*MINIMAL OPERATIONAL CAPABILITY*			
	EMERGENCY POWER | LONG-TERM ALTERNATE POWER SUPPLY -			+
	SELF-CONTAINED			
PE-12	**Emergency Lighting**	x	x	x
PE-13	**Fire Protection**	x	x	x
PE-13 (1)	*FIRE PROTECTION | DETECTION DEVICES/SYSTEMS*			x
PE-13 (2)	*FIRE PROTECTION | SUPPRESSION DEVICES/SYSTEMS*			x
PE-13 (3)	*FIRE PROTECTION | AUTOMATIC FIRE SUPPRESSION*		x	x
PE-14	**Temperature and Humidity Controls**	x	x	x
PE-15	**Water Damage Protection**	x	x	x
PE-15 (1)	*WATER DAMAGE PROTECTION | AUTOMATION SUPPORT*			x
PE-16	**Delivery and Removal**	x	x	x
PE-17	**Alternate Work Site**		x	x
PE-18	**Location of Information Components**			x
PLANNING – PL				
PL-1	**Security Planning Policy and Procedures**	x	x	x
PL-2	**System Security Plan**	x	x	x
PL-2 (3)	*SYSTEM SECURITY PLAN | PLAN/COORDINATE WITH OTHER ORGANIZATIONAL ENTITIES*	+	x	x
PL-4	**Rules of Behavior**	x	x	x
PL-4 (1)	*RULES OF BEHAVIOR | SOCIAL MEDIA AND NETWORKING RESTRICTIONS*		x	x
PL-7	**Security Concept of Operations**		+	+
PL-8	**Information Security Architecture**		x	x
PERSONNEL SECURITY – PS				
PS-1	**Personnel Security Policy and Procedures**	x	x	x
PS-2	**Position Risk Designation**	x	x	x
PS-3	**Personnel Screening**	x	x	x

PS-4	Personnel Termination	x	x	x
PS-4 (2)	*PERSONNEL TERMINATION	AUTOMATED NOTIFICATION*			x
PS-5	Personnel Transfer	x	x	x
PS-6	Access Agreements	x	x	x
PS-7	Third-Party Personnel Security	x	x	x
PS-8	Personnel Sanctions	x	x	x
RISK ASSESSMENT – RA				
RA-1	Risk Assessment Policy and Procedures	x	x	x
RA-2	Security Categorization	x	x	x
RA-3	Risk Assessment	x	x	x
RA-5	Vulnerability Scanning	x	x	x
RA-5 (1)	*VULNERABILITY SCANNING	UPDATE TOOL CAPABILITY*		x	x
RA-5 (2)	*VULNERABILITY SCANNING	UPDATE BY FREQUENCY/PRIOR TO NEW SCAN/WHEN IDENTIFIED*		x	x
RA-5 (4)	*VULNERABILITY SCANNING	DISCOVERABLE INFORMATION*			x
RA-5 (5)	*VULNERABILITY SCANNING	PRIVILEGED ACCESS*		x	x
SYSTEM AND SERVICES ACQUISITION – SA				
SA-1	System and Services Acquisition Policy and Procedures	x	x	x
SA-2	Allocation of Resources	x	x	x
SA-3	System Development Lifecycle	x	x	x
SA-4	Acquisition Process	x	x	x
SA-4 (1)	*ACQUISITION PROCESS	FUNCTIONAL PROPERTIES OF SECURITY CONTROLS*		x	x
SA-4 (2)	*ACQUISITION PROCESS	DESIGN/IMPLEMENTATION INFORMATION FOR SECURITY CONTROLS*		x	x
SA-4 (9)	*ACQUISITION PROCESS	FUNCTIONS/PORTS/PROTOCOLS / SERVICES IN USE*		x	x

SA-4 (10)	*ACQUISITION PROCESS	USE OF APPROVED PIV PRODUCTS*	x	x	x
SA-5	**Information System Documentation**	x	x	x
SA-8	**Security Engineering Principles**		x	x
SA-9	**External Information System Services**	x	x	x
SA-9 (2)	*EXTERNAL INFORMATION SYSTEMS	IDENTIFICATION OF FUNCTIONS / PORTS/PROTOCOLS/SERVICES*		x	x
SA-10	**Developer Configuration Management**		x	x
SA-11	**Developer Security Testing and Evaluation**		x	x
SA-12	**Supply Chain Protection**			x
SA-15	**Development Process, Standards and Tools**			x
SA-16	**Developer-Provided Training**			x
SA-17	**Developer Security Architecture and Design**			x
SYSTEM AND COMMUNICATIONS PROTECTION – SC				
SC-1	**System and Communications Protection Policy and Procedures**	x	x	x
SC-2	**Application Partitioning**		x	x
SC-3	**Security Function Isolation**			x
SC-4	**Information in Shared Resources**		x	x
SC-5	**Denial of Service Protection**	x	x	x
SC-7	**Boundary Protection**	x	x	x
SC-7 (3)	*BOUNDARY PROTECTION	ACCESS POINTS*		x	x
SC-7 (4)	*BOUNDARY PROTECTION	EXTERNAL TELECOMMUNICATIONS SERVICES*		x	x
SC-7 (5)	*BOUNDARY PROTECTION	DENY BY DEFAULT/ALLOW BY EXCEPTION*		x	x
SC-7 (7)	*BOUNDARY PROTECTION	PREVENT SPLIT TUNNELING FOR REMOTE DEVICES*		x	x
SC-7 (8)	*BOUNDARY PROTECTION	ROUTE TRAFFIC TO AUTHENTICATED PROXY SERVERS*			x
SC-7 (18)	*BOUNDARY PROTECTION	FAIL SECURE*		+	**x**

SC-7 (21)	*BOUNDARY PROTECTION	ISOLATION OF INFORMATION SYSTEM COMPONENTS*			x
SC-8	**Transmission Confidentiality and Integrity**		x	x
SC-8 (1)	transmission confidentiality and integrity	cryptographic or alternate physical protection		x	x
SC-10	**Network Disconnect**		x	x
SC-12	**Cryptographic Key Establishment and Management**	x	x	x
SC-12 (1)	*CRYPTOGRAPHIC KEY ESTABLISHMENT AND MANAGEMENT	AVAILABILITY*			x
SC-13	**Cryptographic Protection**	x	x	x
SC-15	**Collaborative Computing Devices**	x	x	x
SC-17	**Public Key Infrastructure Certificates**		x	x
SC-18	**Mobile Code**		x	x
SC-19	**Voice Over Internet Protocol**		x	x
SC-20	**Secure Name/Address Resolution Service (Authoritative Source)**	x	x	x
SC-21	**Secure Name/Address Resolution Service (Recursive or Caching Resolver)**	x	x	x
SC-22	**Architecture and Provisioning for Name/Address Resolution Service**	x	x	x
SC-23	**Session Authenticity**		x	x
SC-24	**Fail in Known State**		+	x
SC-28	**Protection of Information at Rest**		x	x
SC-39	**Process Isolation**	x	x	x
SC-41	**Port and I/O Device Access**	+	+	+
SYSTEM AND INFORMATION INTEGRITY – SI				
SI-1	**System and Information Integrity Policy and Procedures**	x	x	x
SI-2	**Flaw Remediation**	x	x	x
SI-2 (1)	*FLAW REMEDIATION	CENTRAL MANAGEMENT*			x
SI-2 (2)	*FLAW REMEDIATION	AUTOMATED FLAW REMEDIATION STATUS*		x	x

SI-3	**Malicious Code Protection**	x	x	x	
SI-3 (1)	MALICIOUS CODE PROTECTION | CENTRAL MANAGEMENT		x	x	
SI-3 (2)	MALICIOUS CODE PROTECTION | AUTOMATIC UPDATES		x	x	
SI-4	**Information System Monitoring**	x	x	x	
SI-4 (2)	INFORMATION SYSTEM MONITORING | AUTOMATED TOOLS FOR REAL-TIME ANALYSIS		x	x	
SI-4 (4)	INFORMATION SYSTEM MONITORING | INBOUND AND OUTBOUND COMMUNICATIONS TRAFFIC		x	x	
SI-4 (5)	INFORMATION SYSTEM MONITORING | SYSTEM-GENERATED ALERTS		x	x	
SI-5	**Security Alerts, Advisories and Directives**	x	x	x	
SI-5 (1)	SECURITY ALERTS, ADVISORIES AND DIRECTIVES | AUTOMATED ALERTS AND ADVISORIES			x	
SI-6	**Security Function Verification**			x	
SI-7	**Software, Firmware and Information Integrity**		x	x	
SI-7 (1)	SOFTWARE, FIRMWARE AND INFORMATION INTEGRITY | INTEGRITY CHECKS		x	x	
SI-7 (2)	SOFTWARE, FIRMWARE AND INFORMATION INTEGRITY | AUTOMATED NOTIFICATIONS OF INTEGRITY VIOLATIONS			x	
SI-7 (5)	SOFTWARE, FIRMWARE AND INFORMATION INTEGRITY | AUTOMATED RESPONSE TO INTEGRITY VIOLATIONS			x	
SI-7 (7)	SOFTWARE, FIRMWARE AND INFORMATION INTEGRITY | INTEGRATION OF DETECTION AND RESPONSE		x	x	
SI-7 (14)	SOFTWARE, FIRMWARE AND INFORMATION INTEGRITY | BINARY OR MACHINE EXECUTABLE CODE			x	
SI-8	**Spam Protection**		x	x	

SI-8 (1)	SPAM PROTECTION | CENTRAL MANAGEMENT OF PROTECTION MECHANISMS		x	x
SI-8 (2)	SPAM PROTECTION | AUTOMATIC UPDATES		x	x
SI-10	Information Input Validation		x	x
SI-11	Error Handling		x	x
SI-12	Information Handling and Retention	x	x	x
SI-13	Predictable Failure Prevention			+
SI-16	Memory Protection		x	x
SI-17	Fail-Safe Procedures	+	+	+
ORGANIZATION-WIDE INFORMATION SECURITY PROGRAM MANAGEMENT CONTROLS – PM				
PM-1	Information Security Program Plan Policy and Procedures			
PM-2	Senior Information Security Officer			
PM-3	Information Security Resources			
PM-4	Plan of Action and Milestones Process			
PM-5	Information System Inventory			
PM-6	Information Security Measures of Performance			
PM-7	Enterprise Architecture			
PM-8	Critical Infrastructure Plan			
PM-9	Risk Management Strategy			
PM-10	Security Authorization Process			
PM-11	Mission/Business Process Definition			
PM-12	Insider Threat Program			
PM-13	Information Security Workforce			
PM-14	Testing, Training and Monitoring			
PM-15	Contacts with Security Groups and Associations			
PM-16	Threat Awareness Program			

Table A3.1. *NIST SP 800-82 security measures*

Appendix 4

ANSSI Security Measures

This appendix presents the measures proposed by the ANSSI guides (ANSSI 2013a; ANSSI 2013b). They are defined according to the class of the system (Chapter 6).

Recommendations are prefixed with an R and directives with a letter D.

A4.1. Organizational measures

A4.1.1. *Knowledge of the industrial system*

Roles and responsibilities	C1	R1 – A cybersecurity chain of responsibility must be put in place. It should cover all systems. R2 – Responsibilities for cybersecurity should be clearly defined for each of the stakeholders regardless of the aspect concerned (development, integration, operation, maintenance, etc.).
	C2	D3 – R1 is mandatory. D4 – R2 is mandatory.
	C3	D5 – The identity and contact details of the person in charge of the cybersecurity chain of custody must be communicated to the cyber defense authority. D6 – The limits of liability must be reviewed periodically, at least once a year.

Mapping	C1	R7 – Build a map: – physical; – logical (flow); – of applications.
	C2	D8 – Build a map: – physical; – logical (flow); – related applications; – of the system administration. R9 – Review the mapping at least once a year and with each modification.
	C3	D10 – R9 is mandatory.
Risk analysis	C1	R11 – Carry out a risk analysis for cybersecurity, however brief.
	C2	D12 – Carry out a risk analysis for cybersecurity according to a method chosen by the responsible entity.
	C3	D13 – In addition to class 2, updating is required once a year. R14 – Require the analysis to be carried out by a certified service provider.
Backup management	C1	R15 – Set up a backup plan to allow recovery in the event of an incident. R16 – Save the configurations before and after any modification, including when they have been made "hot". R17 – If possible, regularly test the backup process, possibly on a limited but representative subset. The data concerned are all the data needed to restore the system after a disaster: programs, configurations, firmware, process parameters, data useful for traceability, etc.
	C2	D18 – C1's recommendations are mandatory.
	C3	No additional requirements.

Documentation management	C1	In the documentation describing the system (functional analyses, list of variables, various plans, etc.), some documents may be sensitive. R19 – The sensitivity level of the documentation should be defined and appear on the documents that must be processed accordingly. R20 – All design, configuration and operational documents should be considered sensitive. R21 – Store documents in an IS with an appropriate level of sensitivity.
	C2	R22 – Ensure integrity and confidentiality. R23 – Review the documentation at regular intervals.
	C3	D24 – R19, R21 and R22 are mandatory.

Table A4.1. *Recommendations and guidelines for system knowledge*

A4.1.2. *Managing the stakeholders*

Participant management	C1	R25 – Implement stakeholder management procedures (creation/destruction of computer accounts, local access, mobile terminal management, sensitive document management). R26 – Implement a competency management process to ensure that stakeholders have the appropriate competencies.
	C2	D27 – Class 1 recommendations become mandatory.
	C3	D28 – As class 2, plus periodic review at least once a year.
Awareness and training of participants	C1	R29 – Workers should be empowered and trained in cybersecurity. R30 – If possible, set up a charter of good conduct to be signed by all stakeholders.
	C2	D31 – Class 1 recommendations become mandatory.

	C3	D32 – Mandatory cybersecurity training before intervention. R33 – Training provided by a certified service provider. R34 – Provide training at the same time as site safety and security training.
Intervention management	C1	R35 – If possible, set up an intervention management procedure to identify: – intervener and principal; – date and time; – perimeter; – actions taken; – list of equipment removed/replaced; – changes made and impact. R36 – Identify and update the hardware and software used for interventions. R37 – Have the intervention authorization validated by the responsible entity. R38 – Audit the process once a year to verify compliance with the procedure.
	C2	D39 – Class 1 recommendations become mandatory. R40 – If the intervener brings his own tools, in exceptional cases, provide a procedure to verify the level of safety of these tools.
	C3	D41 – As class 2, but the use of special tools brought by the intervener is prohibited.

Table A4.2. *Recommendations and guidelines for participants*

A4.1.3. *Integration of cybersecurity into the lifecycle*

Inclusion in specifications	C1	R42 – Integrate the requirements identified during the specification phases. R43 – Define a contact point in charge of liaising with the responsible entity's chain of responsibility, ensuring compliance with the policy, communicating differences. R44 – Include the expected documents in the specifications: risk analysis, functional analysis, organic analysis, operation and maintenance file, mapping. R45 – Plan cyber security tests during the acceptance phase (follow recommendation R71).
	C2	D46 – C1's recommendations are mandatory. Add to the specifications: D47 – A confidentiality clause. R48 – If possible, a regular clause for revising the risk analysis. R49 – Describe in the specification documents the technical, human and organizational means to trace them and be able to verify the level of cybersecurity. R50 – The contractor must have a security insurance plan describing all measures that meet the requested cybersecurity requirements. R51 – The contractor uses a secure development environment. R52 – The contract includes a clause allowing contractors and suppliers to be audited.
	C3	D53 – C2's recommendations are mandatory. R54 – Add a clause requiring the provision of hardware and software labeled for cybersecurity. R55 – Require software developers to demonstrate that their development processes, employ state-of-the-art engineering methods, quality control processes and validation techniques to reduce software failures and vulnerabilities. R56 – Labeling the contractor.

Integrations in the specification phases	C1	R57 – Take into account all the technical measures presented below. For example, the necessity: – to authenticate participants; – to define a secure architecture; – to secure the equipment; – to be able to requalify a system following security updates. R58 – Provide procedures and technical means to allow preventive and curative maintenance operations in order to maintain the level of cybersecurity over time. R59 – Take into account physical security when defining the location of equipment. R60 – Require in the specifications that operations not essential to the operation of the industrial system be performed on another information system.
	C2	D61 – C1's recommendations are mandatory. R62 – Integrate into the design of tools and mechanisms to manage security and facilitate requirements such as: – configuration control; – the hardening of configurations; – vulnerability management.
	C3	D63 – The recommendations of C2 are mandatory.
Integration in the design phases	C1	R64 – During design, limit interfaces and system complexity to a minimum in order to limit the introduction of vulnerabilities during implementation. Integrate cybersecurity features of equipment (authentication mechanisms, segregation of rights, etc.) into the equipment selection process. R66 – Define roles for integrated stakeholders in the management of computer account rights with a policy of least privilege.
	C2	D67 – C1's recommendations are mandatory.
	C3	No additional requirements.

Appendix 4   335

Cybersecurity audits and tests	C1	R68 – Set up regular audits, which may be internal. R69 – The audit must be followed by an action plan validated and monitored by the responsible entity.
	C2	D70 – C1's recommendations are mandatory. R71 – The audit program must contain: – boundary tests; – error tests of business functions; – testing of verification and exception handling; – the development of threat scenarios (penetration tests and attempted takeovers during maintenance phases); – the verification of security mechanisms (patch deployment, event log analysis, backup recovery, etc.); – the evaluation of the system's performance; – the audits carried out by certified service providers.
	C3	D73 – The elements suggested for the audit in class C2 are mandatory. D74 – Conduct audits at least once a year.
Transfer in operation	C1	R75 – Before commissioning: – establish a comprehensive inventory of the system's cybersecurity level; – ensure that the available means are available to maintain it at an acceptable level.
	C2	D76 – Obligation for the entities responsible to approve the system.
	C3	D77 – Obligation to have the system approved by an external organization. Prior authorization for commissioning is required
Change management	C1	R78 – For PLC programs, SCADA applications and equipment configurations, use tools to quickly check the differences between the current version and the version to be installed, and ensure that only the necessary and requested changes have been applied. R79 – Track changes.

	C2	D80 – C1's recommendations are mandatory. R81 – For PLC programs, SCADA applications, set up a process to check the running program versions against a reference version. R82 – Evaluate changes in a test environment.
	C3	D83 – R78 is mandatory + have the impacts validated by the responsible entity before going into production. D84 – C2's recommendations are mandatory.
Monitoring process	C1	R85 – Implementation of a process to monitor threats and vulnerabilities.
	C2	D86 – R85 is mandatory. R87 – Contractualize the distribution, by suppliers, of vulnerability bulletins for all hardware and software equipment used. R88 – Set up a monitoring process on the evolution of protection techniques (can be based on open sources available such as the ANSSI website).
	C3	D89 – C2's recommendations are mandatory. D90 – Set up a monitoring process on the evolution of attack techniques. In case of significant changes, repeat the risk analysis.
Obsolescence management	C1	R91 – Include clauses in contracts relating to the management of obsolescence of equipment and software (with support end date). Indeed, for software or equipment that is not maintained, vulnerabilities are not fixed. R92 – Include a replacement plan for obsolete equipment and applications.
	C2	D93 – C1's recommendations are mandatory.
	C3	No additional requirements.

Table A4.3. *Recommendations and guidelines for lifecycle management*

A4.1.4. *Physical security of the facilities and premises*

Access to premises	C1	R94 – Provide a physical access control policy, including: – retrieve an employee's keys or badges upon departure; – regularly change the company's alarm codes; – never give keys or alarm codes to external service providers, unless it is possible to trace access and restrict it to specific time slots. R95 – Access to the premises must be logged and auditable.
	C2	D96 – C1's recommendations are mandatory. R97 – The control mechanisms must be robust (ref1). R98 – Put the accesses under video protection. R99 – Restrict access to equipment to authorized persons only.
	C3	D100 – C2's recommendations are mandatory. D101 – Implement an intrusion detection system for vital areas, especially those not occupied 24 h a day.
Access to equipment and wiring	C1	R102 – Install the servers in rooms with access control. R103 – Install station CPUs, PLCs and network equipment in locked cabinets. R104 – Do not leave industrial network access sockets in areas open to the public.
	C2	D105 – C1's recommendations are mandatory. D106 – Do not leave industrial network access sockets in unattended areas. R107 – Protect the physical integrity of the cables. R108 – Close the plugs dedicated to maintenance and define a procedure for removing the plug including prior authorization. R109 – Set up an opening detector with alarm for cabinets of sensitive equipment or visual control means, and define a procedure for opening including prior authorization.
	C3	D110 – C2's recommendations are mandatory.

Table A4.4. *Recommendations and guidelines for physical security*

A4.1.5. *Reaction to the incidents*

A disaster recovery or business continuity plan ensures the recovery or continuity of service following a disaster, regardless of its origin. With regard to cybersecurity, it is based on risk analysis, which has made it possible to identify impacts and threats.

Recovery or business continuity plan	C1	R111 – Set up a backup plan for sensitive data, so that the system can be rebuilt afterwards.
		R112 – Consider cybersecurity in recovery and business continuity plans.
	C2	R113 – Test the plan at least once a year.
	C3	D114 – The recommendations of C1 and C2 are mandatory.
Degraded modes	C1	R115 – Plan to include in the intervention procedures an emergency mode to be able to intervene quickly in case of need, without significantly degrading the level of cybersecurity, in particular not to affect the traceability of the interventions.
		R116 – Integrate degraded modes for stopping allowing them to:
		– either stop without causing damage (material or human);
		– or continue to operate by piloting in "manual" mode.
	C2	No additional requirements.
	C3	D117 – C1's recommendations are mandatory.
Crisis management	C1	R118 – Implement a crisis management process to determine:
		– what to do when an incident is detected;
		– who to alert;

		– who should coordinate actions in the event of a crisis; – what are the first steps to be taken. R119 – Include in the crisis management plan a procedure to manage incidents at the right level of responsibility and decide accordingly: – whether to initiate a disaster recovery plan; – if legal action is necessary. R120 – Include in the crisis management plan a postincident analysis phase to determine the origin of the incident.
	C2	R121 – Test the plan at least once a year.
	C3	D122 – The recommendations of C1 and C2 are mandatory.

Table A4.5. *Recommendations and guidelines for incident response*

A4.2. Technical measures

These technical measures potentially involve:

– servers, workstations and workstations;

– engineering stations and programming consoles;

– mobile equipment: laptops, tablets, computers, etc.;

– supervision software and applications (SCADA);

– CAMM and MES software and applications, if available;

– human–machine interfaces (touch screens);

– PLCs and remote units (RTU);

– network equipment (switches, routers, firewalls, wireless access points);

– intelligent sensors and actuators;

– etc.

A4.2.1. *Authentication of stakeholders*

The authentication of a user in a computer system requires the definition of an account, often associating a name (identifier) and a password (authentication). Other identification modes exist.

Accounts can be associated with an operating system (Windows, Linux, etc.), an application or equipment that uses a dedicated system.

Accounts have different levels of privileges, offering more or fewer possibilities to users. Administrators are often distinguished from simple users.

Account management	C1	R123 – Identify each user in a unique way.
		R124 – Protect all accounts with important privileges such as administrator accounts with an authentication mechanism.
		Separate user and administrator accounts.
		R125 – Avoid generic accounts, especially those with significant privileges; if impossible to avoid, limit them to very specific and documented uses.
		R126 – Define roles, documented and implemented so that users' privileges correspond exactly to their needs.
		R127 – Perform an audit of events related to the use of accounts.
		R128 – Delete or deactivate the accounts of former users.
	C2	D129, R123, R124 and R125 are mandatory.
		Do not use default or generic accounts unless there are strong operational constraints.
		Do not use generic accounts with high privileges.
		Separate high privilege accounts (admin) and user accounts.
		D130, R126 are mandatory with validation by the user's line manager.
		D131, R128 are mandatory and complete D27 (stakeholder management procedure).
		R132 – Conduct an annual review of user accounts to verify the above obligations, with particular attention to administrative accounts.
		R133 – Configure read-only access for first-level maintenance tasks.

		R134 – If account management is centralized, audit the configuration of the centralized directory regularly and at least once a year.
	C3	D135 – R127, R132 and R134 are mandatory.
Authentication management n	C1	R136 – Only make the various components (hardware and software) accessible after authentication with ID and password. R136 – Passwords must be robust, and default passwords must be changed. R137 – Prefer an inhibition delay to blocking in case of authentication failure. R138 – Protect passwords in integrity and confidentiality if transmitted over the network. R139 – If software authentication is not possible due to operational constraints, define and document palliative measures, such as: – apply physical access control; – limit the accessible functionalities; – implement smart card identification; – compartmentalize the equipment more. An example of such a case is a SCADA joint account. R140 – Keep password files, ensuring integrity and availability. R141 – Define a secure procedure to reset passwords in case of loss.
	C2	D142 – C1's recommendations are mandatory. R143 – Use strong authentication (smart card, one time password, etc.) for workstations and servers. Do the same for field equipment (PLCs, remote e/s, etc.) that allow it. R144 – If R143 cannot be applied, reinforce the password policy (keep password history, check its quality, force renewal). R145 – Log authentication failures and successful authentications of privilege accounts.
	C3	D146, R136, R144 and R145 are mandatory. D147, R143 are mandatory for all exposed equipment (workstations, laptops, engineering stations, programming consoles, firewalls, VPN, etc.).

Table A4.6. *Recommendations and guidelines for authentication*

A4.2.2. *Architecture security*

The partitioning of industrial information systems is a key measure to improve security. It is very well developed in the IEC 62443 standard (Chapter 7).

System partitioning	C1	R148 – Cut the system into functional areas or coherent technical areas and partition them. R149 – Implement a filtering policy between zones. R150 – For non-IP flows, filter on source and destination identifiers. It will also be possible to filter on the type of protocol allowed. R151 – As far as possible, achieve physical partitioning between the functional areas of the industrial system. R152 – Perform a logical partitioning during a physical separation by VLAN, for example. R153 – Separate the equipment administration network from other networks (switches, gateways, routers, firewalls, etc.), at a minimum logically. R154 – Keep the administration workstations only for this purpose and do not connect them to the Internet or a management network.
	C2	D155, R148 to R151 are mandatory. D156, R153 and R154 are mandatory. If it is not possible to carry out a partitioning because the equipment does not allow it, study possible countermeasures and assess the level of residual risk. R157 – If possible, make the flows unidirectional between class C1 and C2 systems. A firewall can be used. R158 – Separate the equipment administration network from other networks (physically, at least logically, using VPN tunnels using qualified products).
	C3	D159 – Make unidirectional flows between C3 and non-C3 systems. Unidirectionality is physically ensured by a data diode.

		R160 – Use a labeled diode.
		D161 – Physically separate class 3 systems from lower class systems. It is forbidden to use a logical partitioning.
Interconnection of the management information system (considered by default of class C1)	C1	R162 – Protect the interconnection with a firewall device. R163 – Limit flows to the strict minimum. R164 – Implement a flow filtering policy as above.
	C2	D165 – C1's recommendations are mandatory. R166 – Make the flows unidirectional between the C2 system and the management system. A firewall can be used.
	C3	D167 – Make unidirectional flows between C3 class systems and the management system. Unidirectionality is physically ensured by a data diode. R168 – Use an organism-certified diode.
Internet access and remote sites	C1	R169 – Limit Internet access. In particular, supervisory positions and equipment should not have access to them. R170 – Limit Internet access to the system. R171 – Use an IPSec VPN interconnection between remote sites for confidentiality, integrity and authenticity. R172 – Set up a firewall at the interconnection gateways. R173 – Configure the gateways in a secure way.
	C2	D174, R169, R170, R171 and R172 are mandatory. R175 – Use labeled gateways.
	C3	D176 – Prohibit interconnections between the C3 system and the public network.
Remote access for nomadic workstations	C1	R177 – When remote management, remote maintenance or remote diagnosis operations are required, the following rules should be applied: – make connections at the request of the responsible entity; – authenticate the remote connection equipment;

		– change the login password regularly; – enable logging; – terminate the communication after a maximum period of inactivity; – partition the equipment and filter the flows; – use secure protocols that ensure integrity and authenticity. R178 – In the case of a modem connection, use a callback system. R179 – Use labeled connection equipment.
	C2	D180 – Use a solution with: – a guarantee of confidentiality, integrity and authenticity of communications (example: IPsec VPN); – strong two-factor authentication; – compartmentalize connection equipment and filter flows; – log security events. R181 – Use a detection probe at the gateway to analyze incoming and outgoing traffic.
	C3	D182 – Remote maintenance prohibited (if necessary, integrate remote equipment into the C3 system and apply C3 level rules). D183 – Remote diagnosis possible with the following measures: – remote connection on a partitioned server; – data pushed to the server via a labelled diode.
Distributed industrial systems	C1	R184 – Use secure protocols that protect confidentiality, integrity and authenticity for flows through unprotected networks. R185 – Use VPN labeled gateways at the ends of the links. R186 – Position the equipment behind a firewall allowing only the strictly necessary flows to pass through. In particular, traffic external to the VPN should be blocked. R187 – If necessary, avoid the Internet and use leased lines.

	C2	D188, R185 and R186 are mandatory. R189 – Use a detection probe at the gateway to analyze incoming and outgoing traffic.
	C3	D190 – Prohibit links on public networks. D191 – R189 is mandatory.
Wireless communication	C1	R192 – Limit the use of wireless technologies to what is strictly necessary. R193 – Sign or encrypt and sign flows according to usage. R194 – Use wireless access points with: – authentication of the access point and the device that connects to the infrastructure; – network access control features (e.g. EAP); – connection logging. R195 – Maximize the partitioning of wireless communications by isolating wireless devices in a separate physical or logical network. R196 – If there is no centralized supervision of safety events, review them regularly. R197 – Reduce the range of emissions to a minimum.
	C2	D198, R196 are mandatory. D199 – Regularly apply security patches to wireless network equipment. R200 – Use a detection probe at the connection of the access point with the rest of the network.
	C3	D201, R200 are mandatory. D202 – Avoid the use of wireless links if possible. D203 – Prohibit the use of wireless links if critical availability is required. D204 – Monitor security events centrally. R205 – Use certified equipment.

Protocol security	C1	R206 – Disable unsecured protocols (http, telnet, ftp, etc.) in favor of secure protocols (https, ssh, sftp, etc.) to ensure integrity, confidentiality, authenticity and absence of replay of flows.
	C2	R207 – If protocols cannot be secured for technical and operational reasons, if possible, implement compensatory measures such as: – perimeter protections (firewall); – VPN to ensure the integrity and authenticity of flows.
	C3	D208, R206 and R207 are mandatory.
Hardening of configurations disabling unnecessary components	C1	R209 – For equipment, deactivate: – the default accounts; – unused physical ports; – removable media, if not used; – non-essential services (e.g. web service). R210 – On workstations, laptops and servers, disable or delete: – debugging and development tools for systems in production; – test data and functions, as well as associated accounts; – all non-essential programs. R211 – For PLCs and SCADA applications, disable or delete: – debugging functions (integrators and equipment manufacturers); – mnemonics and comments.
	C2	D212, R209 are mandatory.
	C3	D213, R210 and R211 are mandatory.

Hardening of configurations: reinforcement of protection	C1	R214 – Apply the recommendations for hardening operating systems [1]. R215 – Run applications only with the privileges necessary for their operation.
	C2	D213, R215 are mandatory. R217 – Set up tools for in-depth workstation defense, set up a white list of applications on the equipment that have the right to run. R218 – For PLCs, when the equipment allows it, set up: – access protection to the CPU and/or program; – the restriction of the IP addresses that can be connected; – disabling the remote programming mode.
	C3	D219, R217 and R218 are mandatory. R220 – Use labeled tools.

Table A4.7. *Recommendations and guidelines for architecture*

NOTE.– An antivirus is not always suitable for industrial systems:

– updating signatures often requires a connection to an external system, which can bring vulnerability;

– it may be incompatible with the principles and requirements of operational safety, especially if it is periodically triggered.

A4.2.3. *Equipment security*

In the following section, we distinguish between:

– programming consoles, which are nomadic equipment for PLC programming;

---

[1] Available at: www.ssi.gouv.fr.

– engineering stations, which are fixed workstations dedicated to process engineering of the industrial system;

– administration workstations that are dedicated to the administration of infrastructure equipment (switches, servers, station, firewall, etc.) of the industrial system.

Hardening of configurations: integrity and authenticity	C1	R221 – Integrate an integrity (checksum) and authenticity (signature) verification mechanism for software, updates and configuration files, such as: –firmwares; – standard operating systems and software; – SCADA software packages; – PLC and SCADA programs; – network equipment configuration files.
	C2	D222, R221 are mandatory. R223 – Regularly check the integrity, authenticity of firmware, software and application programs (PLCs, SCADA, etc.). Ideally, once a day automatically.
	C3	D224, D222 are reinforced: The supplier (OEM, developer, integrator, etc.) must sign the critical elements and the responsible entity must verify the signature upon receipt. D225, R223 are mandatory.
Vulnerability management	C1	R226 – Implement a vulnerability management process to: – search for available patches to fix these vulnerabilities; – identify known vulnerabilities and measure their impact on systems; – deploy patches starting with the most important ones; – identify vulnerabilities that could not be fixed (either because of a lack of patches or because the patch could not be applied due to operational constraints). R227 – Apply patches as a priority on the most exposed workstations (workstations, laptops, engineering stations, programming consoles, firewalls, VPN, etc.).
	C2	D228 – Clearly identify uncorrected vulnerabilities, and then implement specific monitoring and palliative measures to reduce exposure due to these vulnerabilities. R229 – Have patches validated by suppliers before deployment.

	C3	R230 – Conduct an effective verification of the application of security patches and possibly use it as a monitoring indicator. D231, R226, R227, R229 and R230 are mandatory R232 – Implement a test environment to ensure that patches do not cause regression.
Connection interfaces: removable media	C1	R233 – Define a policy for the use of removable media (USB key, floppy disk, hard disk, etc.) R234 – As far as possible, limit the use of removable media to the strict minimum. R235 – Use a decontamination station to analyze and decontaminate all removable devices before using them on the industrial system. R236 – Prohibit the connection of undecontaminated, removable devices. R237 – Provide stakeholders with removable media dedicated to industrial systems, and prohibit the use of these media for other uses and the use of other media.
	C2	D238 – C1's recommendations are mandatory. R239 – Disable removable media ports when their use is not necessary. If physical blocking is not possible, deactivate it logically.
	C3	D240, R239 are mandatory. D241 – Set up an airlock to exchange data with the industrial system, located in a controlled area. Data exchange is a one-off action that must be governed by a procedure. R242 – Use a labeled airlock.
Connection interfaces: network access points	C1	R243 – Clearly identify and identify access points. R244 – Disable unused access points (switches, hubs, patchbays, maintenance sockets on fieldbuses, etc.).
	C2	D245 – C1's recommendations are mandatory. D246 – In case of attempted connection and disconnection on network ports, report an alert and process it.
	C3	D247 – Allow network access points to be accessible only in controlled premises.

Connection interfaces: mobile equipment	C1	R248 – Prohibit the use of personal devices (computers, tablets, USB sticks, cameras, etc.). R249 – Set up guidelines for the use of mobile terminals and signage to remind the charter. R250 – Identify and validate the equipment authorized to connect to the systems. R251 – Implement a process for allocating mobile terminals in order to: – validate the assignment of the terminal by the line manager; – ensure traceability between the terminal and its users; – make the user aware of the rules of use in force. R252 – When the equipment contains sensitive data, encrypt its storage memory.
	C2	D253, R248 to R252 are mandatory. R254 – Use equipment dedicated to the industrial system, including for external service providers. R255 – Do not allow this equipment to leave the site.
	C3	D256, R251, R254 and R255 are mandatory.
Security of programming consoles and administration workstations	C1	R257 – Use engineering stations dedicated to engineering tasks, do not connect them to the Internet, install them in premises with access control, apply the hardening rules described above and switch them off when not in use. R258 – Use programming consoles dedicated to maintenance and operation activities, do not connect them to the Internet, do not connect them to systems other than the industrial system, apply the rules for mobile terminals and the rules for hardening configuration and strengthening protection, store them in a secure room, add a clear identifier (visual marking for example). R259 – Use administration workstations dedicated to the administration of infrastructure equipment, do not connect them to the Internet, install them in premises with access control, apply the hardening rules described above and switch them off when not in use. R260 – Do not install development tools on production machines. Only production environments (runtime) should be installed on SCADA servers and stations for example.

		R261 – If the above recommendation is difficult to implement, as in the case of the use of digital control-command systems (DCS), then compensatory solutions should be considered to isolate the system and reduce its attack surface.
	C2	D262, R257 to R261 are mandatory.
	C3	R263 – Do not use administration stations for permanent system monitoring.
Secure development	C1	R264 – Define rules of good programming practice, then apply them and check their implementation. For example, use the advanced options of some compilers or tools dedicated to checking good programming practices.
	C2	R265 – Use a development environment dedicated to the industrial system. R266 – Implement and apply secure coding rules in addition to the good development practices mentioned above. R267 – Systematically use static analysis tools and robustness tests. R268 – Have code audits performed by external service providers.
	C3	D269, R265, R266, R267 and R268 are mandatory. D270 – The security level of the development environment must be verified by audits.

Table A4.8. *Recommendations and guidelines for equipment*

NOTE.– Installation of patches is a tricky operation, because it can jeopardize the proper functioning of the installation. This update should preferably be carried out as part of maintenance operations.

NOTE.– Some compilers, SCADA and PLC development workshops have many options for reporting additional warnings to the user. Often, these options are not enabled by default. They prevent many programming errors and bugs that can lead to vulnerabilities and should be used.

A4.2.4. *Industrial system monitoring*

Event logs	C1	R271 – Define an event management policy to: – determine which relevant events should be taken into account; – organize their storage (volume, shelf life); – define the conditions for analysis (preventive, postincident, etc.); – define the events that should generate alerts. R272 – Enable traceability functions for hardware and software that allow it, such as syslog, SNMPv3, Windows Event. R273 – Implement a centralized and secure event log management system that takes into account backup, confidentiality and integrity. R274 – Trace and record parameter changes for sensors and actuators, servo and control functions, etc. This can be done by some SCADA software.
	C2	D275, R271, R272 and R273 are mandatory. R276 – Regularly analyze newspapers.
	C3	D277, R276 are mandatory. R278 – Implement a SIEM solution centralizing event logs and allowing logs to be correlated to detect security incidents. The SIEM must be placed behind a diode to avoid being considered a class 3 device.
Means of detection	C1	No additional requirements.
	C2	R279 – Implement intrusion detection means on the periphery of the systems and on the points identified as critical, which include in particular: – interconnections between remote systems; – interconnections of remote management systems; – interconnections between the management IS and the industrial IS; – specific connection points to the outside world (e.g. industrial Wi-Fi); – the airlock stations; – the federating network of industrial supervision stations (SCADA); – networks of PLCs considered sensitive.

		R280 – Use labeled detection means.
		R281 – Centralize the events collected by the probes.
		R282 – Develop the process that describes the consideration of events reported by the probes.
	C3	D283, R279, R281 and R282 are obligatory.

Table A4.9. *Recommendations and guidelines for monitoring*

Here is a non-exhaustive list of audit events to configure:

– authentication attempts (success or failure);

– user actions in the system;

– use of privilege accounts;

– failures of safety mechanisms;

– network connection attempts;

– start and stop the audit functionalities;

– activation, deactivation and modification of the behavior or parameters of security mechanisms (authentication, audit generation, etc.);

– actions taken due to a failure in the storage of audits;

– any attempt to export information;

– use of the management function;

– modification of the user group that is part of a role;

– detection of a physical violation;

– any attempt to establish a user session;

– attempts to load, modify or recover programs, firmware or firmware;

– modification of system parameters (time, IP or non-IP address, cycle time, watchdog, etc.);

– modification or forcing of application data;

– switching an equipment into stop, on, stand-by, restart mode.

Appendix 5

Additions to the IEC 62433 Standard

Fundamental requirement	SL1	SL2	SL3	SL4
FR 1 – Identification and authentication control (IAC)				
SR 1.1 – Human user identification and authentication	x	x	x	x
SR 1.1 RE 1 – Unique identification and authentication			x	x
SR 1.1 RE 2 – Multifactor authentication for untrusted networks			x	x
SR 1.1 RE 3 – Multifactor authentication for all networks				x
SR 1.2 – Software process and device identification and authentication		x	x	x
SR 1.2 RE 1 – Unique identification and authentication			x	x
SR 1.3 – Account management	x	x	x	x
SR 1.3 RE 1 – Unified account management			x	x
SR 1.4 – Identifier management	x	x	x	x
SR 1.5 – Authenticator management	x	x	x	x
SR 1.5 RE 1 – Hardware security for software process identity credentials			x	x
SR 1.6 – Wireless access management	x	x	x	x

SR 1.6 RE 1 – Unique identification and authentication			x	x
SR 1.7 – Strength of password-based authentication	x	x	x	x
SR 1.7 RE 1 – Password generation and lifetime restrictions for human users			x	x
SR 1.7 RE 2 – Password lifetime restrictions for all users				x
SR 1.8 – Public key infrastructure certificates		x	x	x
SR 1.9 – Strength of public key authentication		x	x	x
SR 1.9 RE 1 – Hardware security for public key authentication			x	x
SR 1.10 – Authenticator feedback	x	x	x	x
SR 1.11 – Unsuccessful login attempts	x	x	x	x
SR 1.12 – System use notification	x	x	x	x
SR 1.13 – Access via untrusted networks	x	x	x	x
SR 1.13 RE 1 – Explicit access request approval		x	x	x
FR 2 – Use control (UC)				
SR 2.1 – Authorization enforcement	x	x	x	x
SR 2.1 RE 1 – Authorization enforcement for all users		x	x	x
SR 2.1 RE 2 – Permission mapping to roles		x	x	x
SR 2.1 RE 3 – Supervisor override			x	x
SR 2.1 RE 4 – Dual approval				x
SR 2.2 – Wireless use control	x	x	x	x
SR 2.2 RE 1 – Identify and report unauthorized wireless devices			x	x
SR 2.3 – Use control for portable and mobile devices	x	x	x	x
SR 2.3 RE 1 – Enforcement of security status of portable and mobile devices			x	x
SR 2.4 – Mobile code	x	x	x	x
SR 2.4 RE 1 – Mobile code integrity check			x	x
SR 2.5 – Session lock	x	x	x	x
SR 2.6 – Remote session termination		x	x	x
SR 2.7 – Concurrent session control			x	x
SR 2.8 – Auditable events	x	x	x	x
SR 2.8 RE 1 – Centrally managed, system-wide audit trail			x	x

SR 2.9 – Audit storage capacity	x	x	x	x
SR 2.9 RE 1 – Warn when audit record storage capacity threshold reached			x	x
SR 2.10 – Response to audit processing failures	x	x	x	x
SR 2.11 – Timestamps		x	x	x
SR 2.11 RE 1 – Internal time synchronization			x	x
SR 2.11 RE 2 – Protection of time source integrity				x
SR 2.12 – Non-repudiation			x	x
SR 2.12 RE 1 – Non-repudiation for all users				x
FR 3 – System integrity (SI)				
SR 3.1 – Communication integrity	x	x	x	x
SR 3.1 RE 1 – Cryptographic integrity protection			x	x
SR 3.2 – Malicious code protection	x	x	x	x
SR 3.2 RE 1 – Malicious code protection on entry and exit points			x	x
SR 3.2 RE 2 – Central management and reporting for malicious code protection			x	x
SR 3.3 – Security functionality verification	x	x	x	x
SR 3.3 RE 1 – Automated mechanisms for security functionality verification			x	x
SR 3.3 RE 2 – Security functionality verification during normal operation				x
SR 3.4 – Software and information integrity		x	x	x
SR 3.4 RE 1 – Automated notification about integrity violations			x	x
SR 3.5 – Input validation	x	x	x	x
SR 3.6 – Deterministic output	x	x	x	x
SR 3.7 – Error handling		x	x	x
SR 3.8 – Session integrity		x	x	x
SR 3.8 RE 1 – Invalidation of session IDs after session termination			x	x
SR 3.8 RE 2 – Unique session ID generation			x	x
SR 3.8 RE 3 – Randomness of session IDs				x
SR 3.9 – Protection of audit information		x	x	x
SR 3.9 RE 1 – Audit records on write-once media				x
FR 4 – Data confidentiality (DC)				
SR 4.1 – Information confidentiality	x	x	x	x

Requirement				
SR 4.1 RE 1 – Protection of confidentiality at rest or in transit *via* untrusted networks			x	x
SR 4.1 RE 2 – Protection of confidentiality across zone boundaries				x
SR 4.2 – Information persistence		x	x	x
SR 4.2 RE 1 – Purging of shared memory resources			x	x
SR 4.3 – Use of cryptography	x	x	x	x
FR 5 – Restricted data flow (RDF)				
SR 5.1 – Network segmentation	x	x	x	x
SR 5.1 RE 1 – Physical network segmentation		x	x	x
SR 5.1 RE 2 – Independence from non-control system networks			x	x
SR 5.1 RE 3 – Logical and physical isolation of critical networks				x
SR 5.2 – Zone boundary protection	x	x	x	x
SR 5.2 RE 1 – Deny by default, allow by exception		x	x	x
SR 5.2 RE 2 – Island mode			x	x
SR 5.2 RE 3 – Fail close			x	x
SR 5.3 – General purpose person-to-person communication restrictions	x	x	x	x
SR 5.3 RE 1 – Prohibit all general purpose person-to-person communications			x	x
SR 5.4 – Application partitioning	x	x	x	x
FR 6 – Timely response to events (TRE)				
SR 6.1 – Audit log accessibility	x	x	x	x
SR 6.1 RE 1 – Programmatic access to audit logs			x	x
SR 6.2 – Continuous monitoring		x	x	x
FR 7 – Resource availability (RA)				
SR 7.1 – Denial of service protection	x	x	x	x
SR 7.1 RE 1 – Manage communication loads		x	x	x
SR 7.1 RE 2 – Limit DoS effects to other systems or networks			x	x
SR 7.2 – Resource management	x	x	x	x
SR 7.3 – Control system backup	x	x	x	x
SR 7.3 RE 1 – Backup verification		x	x	x
SR 7.3 RE 2 – Backup automation			x	x
SR 7.4 – Control system recovery and reconstitution	x	x	x	x
SR 7.5 – Emergency power	x	x	x	x

SR 7.6 – Network and security configuration settings	x	x	x	x
SR 7.6 RE 1 – Machine-readable reporting of current security settings			x	x
SR 7.7 – Least functionality	x	x	x	x
SR 7.8 – Control system component inventory		x	x	x

Table A5.1. *Definitions of FRs*

Appendix 6

Some Tools

The list of tools used is available at: http://industrialcybersecurity.io/.

Mapping

Grassmarlin: https://github.com/nsacyber/GRASSMARLIN.

Plcscan: https://github.com/meeas/plcscan.

Auditing tools

https://github.com/InteliSecureLabs/Linux_Exploit_Suggester

https://github.com/GDSSecurity/Windows-Exploit-Suggester

Vulnerabilities

Lynis: https://cisofy.com/lynis/.

Nessus: https://www.tenable.com/downloads/nessus.

OpenVAS: http://www.openvas.org/.

IDS and SIEM

Snort: https://www.snort.org/.

Suricata: https://suricata-ids.org/.

Bro: https://www.bro.org/.

Ossec: https://www.ossec.net/.

ELK: https://www.elastic.co/elk-stack.

OSSIM: https://www.alienvault.com/products/ossim.

List of Acronyms and Abbreviations

ANSSI	*Agence nationale de la sécurité des systèmes d'information* [French National Agency for Information Systems Security]
APN	Access Point Name
BCP	Business Continuity Plan
Botnet	Network of computers operating like robots
BSI	British Standards Institution
BPCS	Basic Process Control System
BT	BowTie diagram
CIM	Computer Integrated Manufacturing
COTS	Commercially available Off-The-Shelf
CPS	Cyber-Physical System
CRRF	Cyber Risk Reduction Factor
CSET	CyberSecurity Evaluation Tool
DCOM	Distributed Common Object Model
DCS	Distributed Control Systems
DHS	U.S. Department of Homeland Security
DMZ	DeMilitarized Zone
DNP	Distributed Network Protocol
DoS	Denial-of-Service attack
DRP	Disaster Recovery Plan
EBIOS	*Expression des besoins et identification des objectifs de sécurité* [expression of needs and identification of security objectives]
ENISA	European Union Agency for Network and Information Security
ERP	Enterprise Resource Planning

FDDI	Fiber Distributed Data Interface
FMEA	Failure Modes and Effects Analysis
FMECA	Failure Mode, Effects, and Criticality Analysis
FR	Fundamental Requirement (IEC 62443)
FTP	File Transfer Protocol
GPRS	General Packet Radio Service
HAZOP	Hazard and Operability
HIDS	Host-based Intrusion Detection System
Historian	Database containing temporal evolutions of measures and actions
HMI	Human–Machine Interface
HTTP	HyperText Transfer Protocol
ICMP	Internet Control Message Protocol
ICS	Industrial Control System
ICS-CERT	Industrial Control Systems Cyber Emergency Response Team
IDS	Intrusion Detection System
IEC	International Electrotechnical Commission
IED	Intelligent Electronic Device
IETF	Internet Engineering Task Force
IP	Internet Protocol
IPS	Intrusion Protection System
Ipsec	Internet Protocol Security protocol
IS	Information System
ISMS	Information Security Management System
ISO	International Organization for Standardization
ISS	Information System Security
ISSP	Information Systems Security Policy
IT	Information Technology
LAN	Local Area Network
LPM	*Loi de programmation militaire* [French Military Programming Act]
MAN	Metropolitan Area Network
MES	Manufacturing Execution System
MitM	Man in the Middle
NIDS	Network Intrusion Detection System
NIST	National Institute of Standards and Technology
OPC	Object linking and embedding for Process Control
OS	Operating System

OSI	Open Systems Interconnection model
OT	Operational Technology
PAC	Programmable Automatic Controller
PCI DSS	Payment Card Industry Data Security Standard
PDCA	Plan, Do, Check, Act
PHA	Preliminary Hazard Analysis
PKCS	Public Key Cryptography Standards
PKI	Public Key Infrastructure
PLC	Programmable Logic Controller
PRA	Preliminary Risk Analysis
RTU	Remote Terminal Unit
SAT	Site Acceptance Test
SCADA	Supervisory Control And Data Acquisition system
SIEM	Security Information and Event Management
SIL	Safety Integrity Level
SIS	Safety Instrumented System
SL	Security Level
SOC	Security Operations Center
SoC	System on a Chip
SQL	Structured Query Language
SSH	Secure SHell
TCP	Transmission Control Protocol
VPN	Virtual Private Network
WAN	Wide Area Network
XSS	Cross-site Scripting Attack

References

Abdo, H., Kaouk, M., Flaus, J.-M., and Masse, F. (2018). A safety/security risk analysis approach of Industrial Control Systems: A cyber bowtie – combining new version of attack tree with bowtie analysis. *Computer Security*, 72, 175–195.

Abshier, J. (2004). Securing your control system [Online]. Available at: https://www.controlglobal.com/articles/2004/238 [Accessed July 15, 2018].

ANSSI (2010). Expression des besoins et identification des objectifs de sécurité EBIOS [Online]. Report, Agence nationale de la sécurité des systèmes d'information, Paris.

ANSSI (2012a). Maîtriser la SSI pour les systèmes industriels [Online]. Report, Agence nationale de la sécurité des systèmes d'information, Paris. https://www.ssi.gouv.fr/uploads/2014/01/Managing_Cybe_for_ICS_EN.pdf.

ANSSI (2012b). Recommandations de sécurité relatives aux mots de passe [Online]. Report, Agence nationale de la sécurité des systèmes d'information, Paris. Available at: https://www.ssi.gouv.fr/uploads/IMG/pdf/NP_MDP_NoteTech.pdf.

ANSSI (2013a). Cybersécurité pour les systèmes industriels: Mesures détaillées [Online]. Report, Agence nationale de la sécurité des systèmes d'information, Paris. Available at: https://www.ssi.gouv.fr/uploads/2014/01/industrial_security_WG_detailed_measures.pdf.

ANSSI (2013b). Cybersécurité pour les systèmes industriels: Classification et mesures principales [Online]. Report, Agence nationale de la sécurité des systèmes d'information, Paris. Available at: https://www.ssi.gouv.fr/uploads/2014/01/industrial_security_WG_Classification_Method.pdf.

ANSSI (2014a). Méthode de classification et mesures principales [Online]. Report, Agence nationale de la sécurité des systèmes d'information, Paris. Available at: https://www.ssi.gouv.fr/uploads/2014/01/industrial_security_WG_Classification_Method.pdf.

ANSSI (2014b). Mesures détaillées [Online]. Report, Agence nationale de la sécurité des systèmes d'information, Paris. Available at: https://www.ssi.gouv.fr/uploads/2014/01/industrial_security_WG_detailed_measures.pdf.

ANSSI (2014c). Politique de sécurité des systèmes d'information de l'État [Online]. Report, Agence nationale de la sécurité des systèmes d'information, Paris. Available at: https:// www.ssi.gouv.fr/uploads/IMG/pdf/pssie_anssi.pdf

ANSSI (2018). EBIOS Risk manager [Online]. Report, Agence nationale de la sécurité des systèmes d'information, Paris. Available at: https://www.ssi.gouv.fr/guide/la-methode-ebios-risk-manager-le-guide/ [Accessed October 2018].

ANSSI (n.d.). Certification CSPN [Online]. Agence nationale de la sécurité des systèmes d'information, Paris.

Armstrong, R. and Hunkar, P. (2010). The OPC UA security model for administrators [Online]. OPC Foundation, Scottsdale, AZ 85260-1868 USA.

Bar-El, H. (2010). Introduction to side channel attacks. [Online]. Discretix Technologies. Available at: http://gauss.ececs.uc.edu/Courses/c653/lectures/SideC/intro.pdf.

Basnight, Z., Butts, J., Lopez, J., and Dube, T. (2013). Firmware modification attacks on programmable logic controllers. *International Journal of Critical Infrastructure Protection*, 6, 76–84.

Braband, J. (2017). Towards an IT security risk assessment framework for railway automation. arXiv.

Byres, E.J., Franz, M., and Miller, D. (2004). *The Use of Attack Trees in Assessing Vulnerabilities in SCADA Systems*. International Infrastructure Survivability Workshop, Lisbon, Portugal.

Cárdenas, A.A., Amin, S., Lin, Z.-S., Huang, Y.-L., Huang, C.-Y., and Sastry, S. (2011). Attacks against process control systems: Risk assessment, detection, and response. *ASIACCS*, 11, 355–366.

CMMI (2010). CMMI for Services, Version 1.3. Report. Software Engineering Institute, Pittsburg.

Cole, E. (2017). Defending against the wrong enemy: 2017 SANS insider threat survey. Report, SANS Institute, Swansea.

Conseil européen (2017). Conclusions [Online]. Available at: http://www.consilium.europa.eu/media/21606/19-euco-final-conclusions-fr.pdf.

Cyber-Physical Systems Public Working Group (2017). *Framework for Cyber-Physical Systems: Volume 1, Overview*. NIST Special Publication 1500-201.

Daemen, J. and Rijmen, V. (1999). *The Rijndael Block Cipher: AES Proposal.* Explication du système de Rijndael [Online]. Available at: http://csrc.nist.gov/CryptoToolkit/aes/rijndael/Rijndael.pdf.

Daemen, J. and Rijmen, V.(2002). *The Design of Rijndael*, AES – The Advanced Encryption Standard, Springer-Verlag, 238.

DHS (2007). Recommended practice case study: cross-site scripting. Study, Homeland Security, National Cyber Security Division. Available at: https://ics-cert.us-cert.gov/sites/default/files/recommended_practices/RP_CaseStudy_XSS_20071024_S508C.pdf.

Diffie, W. and Hellman, M.E. (1976). New directions in cryptography. *IEEE Transactions on Information Theory*, 22, 644–654.

Falliere, N., Murchu, L.O., and Chien, E. (2011). W32. Stuxnet Dossier. Symantec-security response, version 1. Report, Symantec. Available at: https://www.symantec.com/content/en/us/enterprise/media/security_response/whitepapers/w32_s tuxnet_dossier.pdf.

Federal Office for Information Security (2016). Industrial control system security— Top 10 threats and countermeasures [Online]. Federal Office for Information Security. Available at: https://www.allianz-fuer-cybersicherheit.de/ACS/DE/_/downloads/BSI-CS_005E.pdf?__blob=publicationFile&v=3

Fernandes, E., Jun, J., and Prakash, A. (2016). Security analysis of emerging smart home applications. *2016 EEE Symposium on Security and Privacy*, 18–37.

Flaus, J.-M. (1994). *La régulation industrielle: régulateurs PID, prédictifs et flous.* Hermès, Paris.

Flaus, J.-M. (2013). *Risk Analysis: Socio-Technical and Industrial Systems.* ISTE Ltd, London, and Wiley, New York.

Flaus, J.-M. and Georgakis, J. (2018a). Machine learning based intrusion detection approaches for industrial IoT control systems: A review. *IoTsm Conference*, London, UK.

Flaus, J.-M. and Georgakis, J. (2018b). Revue des approches pour la détection d'intrusion à base de machine learning pour les systèmes de commande industriels. *Journées C&ESAR 2018*. Rennes, France.

Fovino, I.N., Coletta, A., Carcano, A., and Masera, M. (2012). Critical state-based filtering system for securing SCADA network protocols. *IEEE Transactions on Industrial Electronics*, 59(10), 3943–3950.

F-Secure (n.d.). Brain virus [Online]. Available at: https://www.f-secure.com/v-descs/brain.shtml [Accessed July 15, 2018].

European Union (2012). Directive 2012/18/EU of the European Parliament and of the Council of 4 July 2012 on the control of major-accident hazards involving dangerous substances, amending and subsequently repealing Council Directive 96/82/EC with EEA relevance. Directive, European Union.

European Union (2016a). Regulation (EU) 2016/679 of the European Parliament and of the Council 27 April 2016 on the protection of natural persons with regard to the processing of personal data and on the free movement of such data, and repealing Directive 95/46/EC (General Data Protection Regulation) (Text with EEA relevance). Directive, European Union.

European Union (2016b). Directive (EU) 2016/1148 of the European Parliament and of the Council of 6 July 2016 concerning measures for a high common level of security of network and information systems across the Union. Directive, European Union.

Goble, W.M. and Cheddie, H. (2005). *Safety Instrumented Systems Verification: Practical Probabilistic Calculations*. ISA, Durham, NC.

Govil, N., Agrawal, A., and Tippenhauer, N.O. (2018). On ladder logic bombs in industrial control systems. In *Computer Security*, Katsikas S. *et al.* (eds). Springer.

Hadziosmanovic, D., Sommer, R., Zambon, E., and Hartel, P.H. (2013). Through the eye of the PLC: Towards semantic security monitoring for industrial control systems. *ACSAC 14: Proceedings of the 30th Annual Computer Security Applications Conference*, New Orleans, LA, 126–135.

Hanna, S., Kumar, S., and Weber, D. (2018). IIC endpoint security best practices. Guide d'usage, Industrial Internet Consortium.

Hauet J.P. (2012). ISA99/IEC 62443: A solution to cybersecurity issues? *ISA Automation Conference*. Doha, Qatar.

Hauet, J.P. (n.d.a). La norme ISA/IEC 62443 (ISA-99) et la cybersécurité des systèmes de contrôle. Course, ISA France.

Hauet J.P. (n.d.b). L'Internet industriel des objets: Les futures architectures de systèmes d'automatisme et de contrôle. Course, ISA France.

Hei, X., Du, X., Lin, S., and Lee, I. (2013). PIPAC: Patient infusion pattern based access control scheme for wireless insulin pump system. *2013 Proceedings IEEE INFPCOM*. Turin, Italy.

High-Tech Bridge Security Research (2016). 90% of SSL VPNs use insecure or outdated encryption, putting your data at risk [Online]. Available at: https://www.htbridge.com/blog/90-percent-of-ssl-vpns-use-insecure-or-outdated-encryption.html [Accessed November 16, 2018].

Howard, M. and Leblanc, D.E. (2002). *Writing Secure Code*. Microsoft Press, Redmond.

Hutchins, E.M., Cloppert, M.J., and Amin, R.M. (2011). Intelligence-driven computer network defense informed by analysis of adversary campaigns and intrusion kill chains. White paper, Lockheed Martin Corporation.

Idaho National Laboratory (n.d.). Aurora generator test [Online]. Available at: https://www.youtube.com/watch?v=fJyWngDco3g [Accessed July 15, 2018].

IEC (2010). Functional safety of electrical/electronic/programmable electronic safety-related systems. IEC 61508:2010.

IEC (2011). Information technology–Security techniques–Information security risk management. ISO/IEC 27005:2011.

IEC (2016). Functional safety–Safety instrumented systems for the process industry sector. IEC 61511:2016.

IEC (n.d.). EDSA Certification [Online]. IEC 62443. Available at: https://www.isasecure.org/en-US/Certification/IEC-62443-EDSA-Certification. [Accessed November 16, 2018].

IEEE (2013). IEEE Standard for intelligent electronic devices cyber security capabilities. IEEE Std 1686TM-2013. IEEE Power Energy Society.

Illera, A.G. and Vidal, J.V. (2014). *Lights Off! The Darkness of the Smart Meters*. Black Hat Europe, Amsterdam, The Netherlands.

International Atomic Energy Agency (2011). Computer security at nuclear facilities. Report, 17, IAEA Nuclear Security Series, International Atomic Energy Agency.

International Atomic Energy Agency (2015). Computer security of instrumentation and control systems at nuclear facilities. Report, 33-T, IAEA Nuclear Security Series, International Atomic Energy Agency.

International Atomic Energy Agency (2016). Conducting computer security assessments at nuclear facilities, Report, IAEA Nuclear Security Series, International Atomic Energy Agency.

ISA (2018). The 62443 series of standards, industrial automation and control systems security. Collection of standards. ISA.

ISO/IEC (2013). Information technology–Security techniques–Code of practice for information security management. Practical manual. ISO/IEC 27002:2013.

Jenkins, G. (2014). Risk methodology. Information security framework programme. Risk Assessment Document, Cardiff University, Cardiff.

Joint Task Force Transformation Initiative Interagency Working Group (2012). Guide for conducting risk assessments. NIST SP 800-30. Guide, National Institute of Standards and Technology, Gaithersburg.

Joye, M. and Olivier, F. (2011). Side-channel analysis. *Encyclopedia of Cryptography and Security*. Springer, Basel.

Kaspersky Lab ICS CERT (2018). Threat landscape for industrial automation systems in H2 2017 [Online]. Available at: https://ics-cert.kaspersky.com/ reports/2018/03/26/threat-landscape-for-industrial-automation-systems-in-h2-2017/ [Accessed July 15, 2018].

Knapp, E.D. and Thomas, J. (2015). *Industrial Network Security: Securing Critical Infrastructure Networks for Smart Grid, SCADA, and Other Industrial Control Systems*. Elsevier, Amsterdam.

Kocher, P.C. (1996). Timing attacks on implementations of Diffie-Hellman, RSA, DSS, and other systems. In *Lecture Notes in Computer Science*, Koblitz N. (ed.). 104–113, Springer.

Kordy, B., Mauw, S., Radomirović, S., and Schweitzer, P. (2011). Foundations of attack–defense trees. In *Formal Aspects of Security and Trust*, Degano P., Etalle S., Guttman J. (eds). Springer, Berlin, 80–95.

Kovacs, E. (2018). Severe flaws expose Moxa industrial routers to attacks [Online]. Available at: https://www.securityweek.com/severe-flaws-expose-moxa-industrial-routers-attacks [Accessed July 2018].

Kriaa, S., Pietre-Cambacedes, L., Bouissou, M., and Halgand, Y. (2015). A survey of approaches combining safety and security for industrial control systems. *Reliability Engineering System Safety*, 139, 156–178.

Legifrance (2016). Annexe. *JORF*. 0145.

Legifrance (2018a). (201AD) Dispositions tendant à transposer la directive (UE) 2016/1148 du parlement européen et du conseil du 6 juillet 2016 concernant des mesures destinées à assurer un niveau élevé commun de sécurité des réseaux et des systèmes d'information dans l'union. Loi. 2018-133.

Legifrance (2018b). Chapitre IV: Dispositions relatives à la protection des infrastructures vitales contre la cyber-menace. Dans Loi no. 2013-1168 du 18 décembre 2013 relative à la programmation militaire pour les années 2014 à 2019 et portant diverses dispositions concernant la défense et la sécurité nationale [Online]. Available at: https://www.legifrance.gouv.fr/affichTexte.do?cidTexte=JORFTEXT000028338825.

Lévy-Bencheton, C., Marinos, L., Mattioli, R., King, T., and Christoph Dietzel, J.S. (2015). Threat landscape and good practice guide for Internet infrastructure. Study, European Union Agency for Network and Information Security.

Li, X., Jiang, P., Chen, T., Luo, X., and Wen, Q. (2017). A survey on the security of blockchain systems. Future generation computer systems [Online]. Available at: https://doi.org/10.1016/j.future.2017.08.020.

Lin, S.-W., Miller, B., Durand, J., Bleakley, G., Chigani, A., Martin, R., Murphy, B., and Crawford, M. (2017). The industrial Internet of Things volume G1: Reference architecture. Report, Industrial Internet Consortium.

Linux-Foundation (2017). Blockchain technologies for business [Online]. Available at: https://www.hyperledger.org/.

Liu, Y., Ning, P., and Reiter, M.K. (2011). False data injection attacks against state estimation in electric power grids CCS '09. *Proceedings of the 16th ACM conference on computer and communications security*, Chicago, USA.

Louis, M., Adrian, B., and Evangelos, R. (2016). ENISA threat landscape 2015. Report, European Union Agency for Network and Information Security.

Macaulay, T. and Singer, B. (2012). *Cybersecurity for Industrial Control Systems*. CRC Press, Boca Raton, FL.

Mahnke, W., Leitner, S., and Damm, M. (2009). *OPC Unified Architecture*. Springer, Berlin.

Manadhata, P.K. and Wing, J.M. (2011). An attack surface metric. *IEEE Transactions on Software Engineering*, 37(3), 371–386.

Mangard, S., Oswald, E., and Popp, T. (2007). *Power Analysis Attack*. Springer, New York.

Mathew, K., Tabassum, M., and Lu Ai Siok, M.V. (2014). A study of open ports as security vulnerabilities in common user computers. *International Conference on Computational Science and Technology (ICCST)*, Kota Kinabalo, Malaysia.

Mauw, S. and Oostdijk, M. (2006). Foundations of attack trees. In *Information Security and Cryptology – ICISC 2005*, Won, D.H., Kim, S. (eds). Springer, Berlin, 186–198.

May, I., David, J., Cohen, F., and Marietta, M. (2018). One year after WannaCry: Assessing the aftermath. *Network Security*, 2018(5), 1–2.

McQueen, M.A., Boyer, W.F., Flynn, M.A., and Beitel, G.A. (2005). Quantitative cyber risk reduction estimation for a SCADA control system. *Proceedings of the 39th Annual Hawaii International Conference on System Sciences (HICSS'06)*, Kauia, USA.

Meserve, J. (2007). Mouse click could plunge city into darkness [Online]. CNN. Available at: http://www.cnn.com/2007/US/09/27/power.at.risk/index.html [Accessed July15, 2018].

Miller, C. and Valasek, C. (2015). Remote exploitation of an unaltered passenger vehicle. Research paper.

Minerva, R., Biru, A., and Rotondi, D. (2015). Towards a definition of the Internet of Things. White paper, IEEE IoT Initiative.

National Cyber Security Centre (2017). TRITON malware targeting safety controllers. Report, National Cyber Security Centre.

van Niekerk, J., von Solms, R. (2016). From information security to cyber security. *Computers & Security*, 38, 97–102.

NIST (2014). Security and privacy controls for federal information systems and organizations. Report, 800-53Ar4, National Institute of Standards and Technology, Gaithersburg.

NIST (2018). Framework for improving critical infrastructure cybersecurity [Online]. Report, National Institute of Standards and Technology. Available at: https://nvlpubs.nist.gov/nistpubs/CSWP/ NIST.CSWP.04162018.pdf.

NSA (2017). GRASSMARLIN user guide. Guide, National Security Agency.

PA Consulting Group (2015). Manage industrial control systems lifecycle: A good practice guide. National Cyber Security Centre.

Pettey, C. (2017). When IT and operational technology converge [Online]. Smarter with Gartner. Available at: https://www.gartner.com/smarterwithgartner/when-it-and-operational-technology-converge/ [Accessed July 15, 2018].

Pollet, J. (2010). Electricity for free? The dirty underbelly of SCADA and smart meters. *Proceedings of Black Hat USA*, Las Vegas, USA.

Raymond, D.R., Marchany, R.C., Brownfield, M.I., and Midkiff, S.F. (2009). Effects of denial-of-sleep attacks on wireless sensor network MAC protocols. *IEEE Transactions on Vehicular Technology*, 58(1), 367–380.

Rivest, R.L., Shamir, A., and Adleman, L. (1978). A method for obtaining digital signatures and public-key cryptosystems. *Communications of the ACM*, 21(1), 120–126.

Ronen, E., Shamir, A., Weingarten, A., and O'Flynn, C. (2017). IoT goes nuclear: Creating a Zigbee chain reaction. *2017 IEEE Symposium on Security and Privacy (SP)*, San Jose, USA.

Schneier, B. (1999). Attack trees. *Dr. Dobb's Journal of Software Tools*, 24(12), 21–29.

Schuett, C., Butts, J., and Dunlap, S. (2014). An evaluation of modification attacks on programmable logic controllers. *International Journal of Critical Infrastructure Protection*, 7(1), 61–68.

Security Now (2018). Schneider electric offers additional details on Triton malware [Online]. Available at: https://www.securitynow.com/author.asp?section_id=613&doc_id=739868 [Accessed July 15, 2018].

Stouffer, K. *et al.* (2017). Cybersecurity framework manufacturing profile [Online]. Report, National Institute of Standards and Technology. Available at: https://nvlpubs.nist.gov/nistpubs/ir/2017/nist.ir.8183.pdf.

Stouffer, K.A., Pillitteri, V.Y., Lightman, S., Abrams, M., and Hahn, A. (2015). Guide to industrial control systems (ICS). SP 800-82, Guide, National Institute of Standards and Technology.

Symantec (2017). Ransom.Wannacry [Online]. Available at: https://www.symantec.com/security-center/writeup/2017-051310-3522-99 [Accessed July15, 2018].

The-Bitcoin-Foundation (2014). How does bitcoin work? [Online]. Available at: https://bitcoin.org/en/how-it-works.

The German Federal Office for Information Security (2011). Threats catalogue [Online].

TÜV SÜD (n.d.). Certification acc. to IEC 62443 [Online]. Available at: https://www.tuev-sued.de/topics/information-technology-it/industrial-it-security/certification-acc.-to-iec-62443.

Université Grenoble-Alpes (2018). Système de sécurisation de procédé cyber-physique. Brevet, FR18/53618.

Voas, J. (2016). Networks of "Things". SP.800-183. Special edition, National Institute of Standards and Technology.

Ware, B. *et al.* (2017). Insider attacks industry survey. Study, Haystax Technology.

Williams, T.J. (1994). The Purdue enterprise reference architecture. *Computers in Industry*, 24(2–3), 141–158.

Wooldridge, S. (2005). SCADA/business network separation: Securing an integrated SCADA system [Online]. Electric Energy Online. Available at: http://www.electricenergyonline.com/energy/magazine/211/article/SCADA-Business.htm [Accessed 15 July, 2018].

Yang, K., Hicks, M., Dong, Q., Austin, T., and Sylvester, D. (2016). A2: Analog malicious hardware. *37th IEEE Symposium on Security and Privacy*, San Jose, USA.

Zurawski, R. (2014). *The Industrial Communication Technology Handbook*. CRC Press, Boca Raton, FL.

Index

6LowPAN, 37
802.15.4, 37

A

accountability, 59
AIC criteria, 57
AMQP, 54
anomaly, 266
architecture security, 253
ARP, 42
asset, 70
attack(s), 91, 71
   surface, 124
   vector, 91, 98
Aurora, 114

B

backdoor, 101
BCPS, 8
blue team, 138
botnet, 102
bowtie diagram, 211
brute force, 104
buffer overflow, 103

C

Censys, 139
CHAZOP, 243
CIP, 50
CMMI, 182
CoAP, 54
conduit, 182
confidentiality, 59
ControlNet, 50
Converged Plantwide EtherNet, 27
CSET, 130
cyber-APR, 239
cyber
   bowtie diagram, 245
   HAZOP, 243
   risk reduction factor, 187

D

data-diode, 260
DCS, 5
DDoS, 102
defense in depth, 80
dematerialized zone, 254
denial of service, 110
DeviceNet, 50
differences between IT and OT
   systems, 64
directive
   NIS, 164
   Seveso, 163
DMZ, 254
DNP3, 48

E

EBIOS, 221
Ethernet, 35
Ethernet/IP, 50
EtherNet/IP, 51

F, G, H

Firewall, 20
FMEA, 206
functional safety, 200
GDPR, 166
good practices, 276
GRAFCET, 11
HART, 48
HAZOP, 207
HIDS, 263
HMI, 15

I

IACS, 169
ICMP, 42
IEC
  27000, 142
  27002, 220
  60870, 48
  61508, 200
  61511, 201
  62443, 170
IED, 13
IIoT, 5, 30
impact, 70
Industrial
  control system, 2
  internet consortium, 159
internal fraud, 114
inventory, 249
IoT platform, 21
IP, 39
ISMS, 82
ISSP, 87
IT, 4
IT/OT convergence, 68

L, M

Ladder diagram, 11
LAN, 33
level(s)
  of protection, 183
life cycle, 22
logical bomb, 101
LOPA, 208
LPM, 163
lSA100.11a, 37
LTE, 38
Man-in-the-Middle, 109
mapping, 249
Maturity Level, 182
Modbus, 45
MQTT, 53

N, O, P

NIDS, 262
NIST Cyber Security Framework, 273
NIST Framework, 145
non-repudiation, 59
OIV, 163
OPC, 51
OPC-UA, 52
OSE, 164
OSI, 34
OT, 4
PAC, 13
PDCA, 83
pentest, 137
PHA, 204
Philips Hue lamps, 116
phishing, 113
PLC, 8
port scan, 111
Profibus, 46
protection level, 183
purdue model, 26

R

ransomware, 103
red team, 138
risk, 73
rootkit, 101
router, 20
RTU, 12

S

SCADA, 4
scenario, 74
secure
  boot, 161
  element, 161
security program elements, 189
Shodan, 139
SIEM, 268
signature, 265
SIL, 199
SIS, 200
SL, 174
SL-T, 177
SmartContracts, 305
spyware, 101
SQL injection, 108
STRIDE, 216
Stuxnet, 115
Switch, 19

SYN, 40
  flood, 41
system
  class, 155
  cyber-physical, 1
  information, 1
  integrators, 170
  security management, 82

T, U, V

TCP, 39
threat, 71
tree
  attack, 234
  fault, 210
Triton, 116
Trojan horse, 100
UDP, 42
virus, 100
VPN, 34
vulnerabilty/vulnerabilities, 128
VxWorks, 12

W, X, Z

WiFi, 36
XMPP, 55
zero day, 105

Other titles from

in

Systems and Industrial Engineering – Robotics

2019

BRIFFAUT Jean-Pierre
From Complexity in the Natural Sciences to Complexity in Operations Management Systems
(Systems of Systems Complexity Set – Volume 1)

2018

BERRAH Lamia, CLIVILLÉ Vincent, FOULLOY Laurent
Industrial Objectives and Industrial Performance: Concepts and Fuzzy Handling

GONZALEZ-FELIU Jesus
Sustainable Urban Logistics: Planning and Evaluation

GROUS Ammar
Applied Mechanical Design

LEROY Alain
Production Availability and Reliability: Use in the Oil and Gas Industry

MARÉ Jean-Charles
Aerospace Actuators 3: European Commercial Aircraft and Tiltrotor Aircraft

MAXA Jean-Aimé, BEN MAHMOUD Mohamed Slim, LARRIEU Nicolas
Model-driven Development for Embedded Software: Application to Communications for Drone Swarm

MBIHI Jean
Analog Automation and Digital Feedback Control Techniques
Advanced Techniques and Technology of Computer-Aided Feedback Control

MORANA Joëlle
Logistics

SIMON Christophe, WEBER Philippe, SALLAK Mohamed
Data Uncertainty and Important Measures
(Systems Dependability Assessment Set – Volume 3)

TANIGUCHI Eiichi, THOMPSON Russell G.
City Logistics 1: New Opportunities and Challenges
City Logistics 2: Modeling and Planning Initiatives
City Logistics 3: Towards Sustainable and Liveable Cities

ZELM Martin, JAEKEL Frank-Walter, DOUMEINGTS Guy, WOLLSCHLAEGER Martin
Enterprise Interoperability: Smart Services and Business Impact of Enterprise Interoperability

2017

ANDRÉ Jean-Claude
From Additive Manufacturing to 3D/4D Printing 1: From Concepts to Achievements
From Additive Manufacturing to 3D/4D Printing 2: Current Techniques, Improvements and their Limitations
From Additive Manufacturing to 3D/4D Printing 3: Breakthrough Innovations: Programmable Material, 4D Printing and Bio-printing

ARCHIMÈDE Bernard, VALLESPIR Bruno
Enterprise Interoperability: INTEROP-PGSO Vision

CAMMAN Christelle, FIORE Claude, LIVOLSI Laurent, QUERRO Pascal
Supply Chain Management and Business Performance: The VASC Model

FEYEL Philippe
Robust Control, Optimization with Metaheuristics

MARÉ Jean-Charles
Aerospace Actuators 2: Signal-by-Wire and Power-by-Wire

POPESCU Dumitru, AMIRA Gharbi, STEFANOIU Dan, BORNE Pierre
Process Control Design for Industrial Applications

RÉVEILLAC Jean-Michel
Modeling and Simulation of Logistics Flows 1: Theory and Fundamentals
Modeling and Simulation of Logistics Flows 2: Dashboards, Traffic Planning and Management
Modeling and Simulation of Logistics Flows 3: Discrete and Continuous Flows in 2D/3D

2016

ANDRÉ Michel, SAMARAS Zissis
Energy and Environment
(Research for Innovative Transports Set - Volume 1)

AUBRY Jean-François, BRINZEI Nicolae, MAZOUNI Mohammed-Habib
Systems Dependability Assessment: Benefits of Petri Net Models (Systems Dependability Assessment Set - Volume 1)

BLANQUART Corinne, CLAUSEN Uwe, JACOB Bernard
Towards Innovative Freight and Logistics (Research for Innovative Transports Set - Volume 2)

COHEN Simon, YANNIS George
Traffic Management (Research for Innovative Transports Set - Volume 3)

MARÉ Jean-Charles
Aerospace Actuators 1: Needs, Reliability and Hydraulic Power Solutions

REZG Nidhal, HAJEJ Zied, BOSCHIAN-CAMPANER Valerio
Production and Maintenance Optimization Problems: Logistic Constraints and Leasing Warranty Services

TORRENTI Jean-Michel, LA TORRE Francesca
Materials and Infrastructures 1 (Research for Innovative Transports Set - Volume 5A)
Materials and Infrastructures 2 (Research for Innovative Transports Set - Volume 5B)

WEBER Philippe, SIMON Christophe
Benefits of Bayesian Network Models
(Systems Dependability Assessment Set – Volume 2)

YANNIS George, COHEN Simon
Traffic Safety (Research for Innovative Transports Set - Volume 4)

2015

AUBRY Jean-François, BRINZEI Nicolae
Systems Dependability Assessment: Modeling with Graphs and Finite State Automata

BOULANGER Jean-Louis
CENELEC 50128 and IEC 62279 Standards

BRIFFAUT Jean-Pierre
E-Enabled Operations Management

MISSIKOFF Michele, CANDUCCI Massimo, MAIDEN Neil
Enterprise Innovation

2014

CHETTO Maryline
Real-time Systems Scheduling
Volume 1 – Fundamentals
Volume 2 – Focuses

DAVIM J. Paulo
Machinability of Advanced Materials

ESTAMPE Dominique
Supply Chain Performance and Evaluation Models

FAVRE Bernard
Introduction to Sustainable Transports

GAUTHIER Michaël, ANDREFF Nicolas, DOMBRE Etienne
Intracorporeal Robotics: From Milliscale to Nanoscale

MICOUIN Patrice
Model Based Systems Engineering: Fundamentals and Methods

MILLOT Patrick
Designing Human–Machine Cooperation Systems

NI Zhenjiang, PACORET Céline, BENOSMAN Ryad, RÉGNIER Stéphane
Haptic Feedback Teleoperation of Optical Tweezers

OUSTALOUP Alain
Diversity and Non-integer Differentiation for System Dynamics

REZG Nidhal, DELLAGI Sofien, KHATAD Abdelhakim
Joint Optimization of Maintenance and Production Policies

STEFANOIU Dan, BORNE Pierre, POPESCU Dumitru, FILIP Florin Gh., EL KAMEL Abdelkader
Optimization in Engineering Sciences: Metaheuristics, Stochastic Methods and Decision Support

2013

ALAZARD Daniel
Reverse Engineering in Control Design

ARIOUI Hichem, NEHAOUA Lamri
Driving Simulation

CHADLI Mohammed, COPPIER Hervé
Command-control for Real-time Systems

DAAFOUZ Jamal, TARBOURIECH Sophie, SIGALOTTI Mario
Hybrid Systems with Constraints

FEYEL Philippe
Loop-shaping Robust Control

FLAUS Jean-Marie
Risk Analysis: Socio-technical and Industrial Systems

FRIBOURG Laurent, SOULAT Romain
Control of Switching Systems by Invariance Analysis: Application to Power Electronics

GROSSARD Mathieu, REGNIER Stéphane, CHAILLET Nicolas
Flexible Robotics: Applications to Multiscale Manipulations

GRUNN Emmanuel, PHAM Anh Tuan
Modeling of Complex Systems: Application to Aeronautical Dynamics

HABIB Maki K., DAVIM J. Paulo
Interdisciplinary Mechatronics: Engineering Science and Research Development

HAMMADI Slim, KSOURI Mekki
Multimodal Transport Systems

JARBOUI Bassem, SIARRY Patrick, TEGHEM Jacques
Metaheuristics for Production Scheduling

KIRILLOV Oleg N., PELINOVSKY Dmitry E.
Nonlinear Physical Systems

LE Vu Tuan Hieu, STOICA Cristina, ALAMO Teodoro, CAMACHO Eduardo F., DUMUR Didier
Zonotopes: From Guaranteed State-estimation to Control

MACHADO Carolina, DAVIM J. Paulo
Management and Engineering Innovation

MORANA Joëlle
Sustainable Supply Chain Management

SANDOU Guillaume
Metaheuristic Optimization for the Design of Automatic Control Laws

STOICAN Florin, OLARU Sorin
Set-theoretic Fault Detection in Multisensor Systems

2012

AÏT-KADI Daoud, CHOUINARD Marc, MARCOTTE Suzanne, RIOPEL Diane
Sustainable Reverse Logistics Network: Engineering and Management

BORNE Pierre, POPESCU Dumitru, FILIP Florin G., STEFANOIU Dan
Optimization in Engineering Sciences: Exact Methods

CHADLI Mohammed, BORNE Pierre
Multiple Models Approach in Automation: Takagi-Sugeno Fuzzy Systems

DAVIM J. Paulo
Lasers in Manufacturing

DECLERCK Philippe
Discrete Event Systems in Dioid Algebra and Conventional Algebra

DOUMIATI Moustapha, CHARARA Ali, VICTORINO Alessandro, LECHNER Daniel
Vehicle Dynamics Estimation using Kalman Filtering: Experimental Validation

GUERRERO José A, LOZANO Rogelio
Flight Formation Control

HAMMADI Slim, KSOURI Mekki
Advanced Mobility and Transport Engineering

MAILLARD Pierre
Competitive Quality Strategies

MATTA Nada, VANDENBOOMGAERDE Yves, ARLAT Jean
Supervision and Safety of Complex Systems

POLER Raul *et al.*
Intelligent Non-hierarchical Manufacturing Networks

TROCCAZ Jocelyne
Medical Robotics

YALAOUI Alice, CHEHADE Hicham, YALAOUI Farouk, AMODEO Lionel
Optimization of Logistics

ZELM Martin *et al.*
Enterprise Interoperability –I-EASA12 Proceedings

2011

CANTOT Pascal, LUZEAUX Dominique
Simulation and Modeling of Systems of Systems

DAVIM J. Paulo
Mechatronics

DAVIM J. Paulo
Wood Machining

GROUS Ammar
Applied Metrology for Manufacturing Engineering

KOLSKI Christophe
Human–Computer Interactions in Transport

LUZEAUX Dominique, RUAULT Jean-René, WIPPLER Jean-Luc
Complex Systems and Systems of Systems Engineering

ZELM Martin, *et al.*
Enterprise Interoperability: IWEI2011 Proceedings

2010

BOTTA-GENOULAZ Valérie, CAMPAGNE Jean-Pierre, LLERENA Daniel, PELLEGRIN Claude
Supply Chain Performance / Collaboration, Alignement and Coordination

BOURLÈS Henri, GODFREY K.C. Kwan
Linear Systems

BOURRIÈRES Jean-Paul
Proceedings of CEISIE'09

CHAILLET Nicolas, REGNIER Stéphane
Microrobotics for Micromanipulation

DAVIM J. Paulo
Sustainable Manufacturing

GIORDANO Max, MATHIEU Luc, VILLENEUVE François
Product Life-Cycle Management / Geometric Variations

LOZANO Rogelio
Unmanned Aerial Vehicles / Embedded Control

LUZEAUX Dominique, RUAULT Jean-René
Systems of Systems

VILLENEUVE François, MATHIEU Luc
Geometric Tolerancing of Products

2009

DIAZ Michel
Petri Nets / Fundamental Models, Verification and Applications

OZEL Tugrul, DAVIM J. Paulo
Intelligent Machining

PITRAT Jacques
Artificial Beings

2008

ARTIGUES Christian, DEMASSEY Sophie, NERON Emmanuel
Resources–Constrained Project Scheduling

BILLAUT Jean-Charles, MOUKRIM Aziz, SANLAVILLE Eric
Flexibility and Robustness in Scheduling

DOCHAIN Denis
Bioprocess Control

LOPEZ Pierre, ROUBELLAT François
Production Scheduling

THIERRY Caroline, THOMAS André, BEL Gérard
Supply Chain Simulation and Management

2007

DE LARMINAT Philippe
Analysis and Control of Linear Systems

DOMBRE Etienne, KHALIL Wisama
Robot Manipulators

LAMNABHI Françoise *et al.*
Taming Heterogeneity and Complexity of Embedded Control

LIMNIOS Nikolaos
Fault Trees

2006

FRENCH COLLEGE OF METROLOGY
Metrology in Industry

NAJIM Kaddour
Control of Continuous Linear Systems